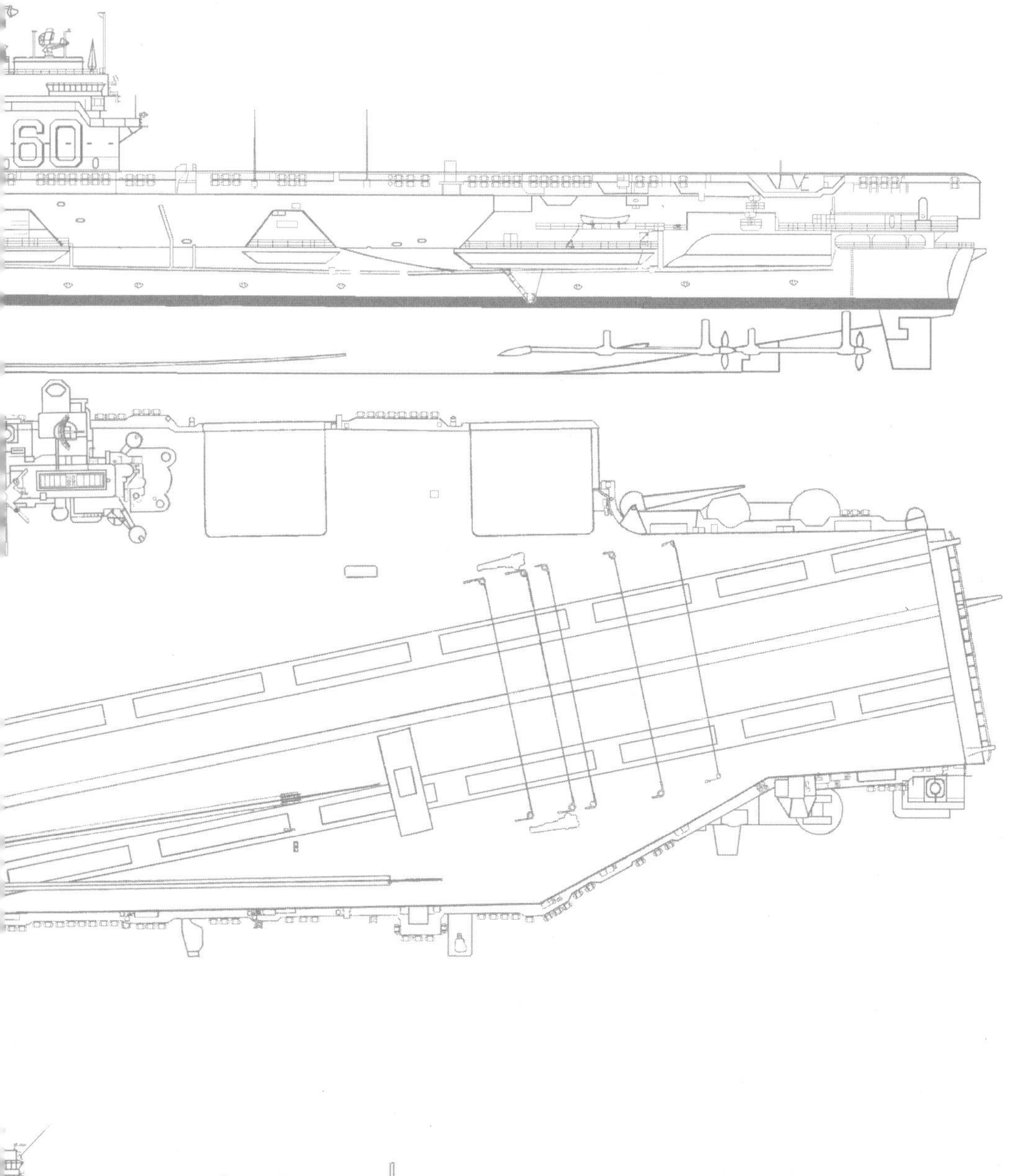
60

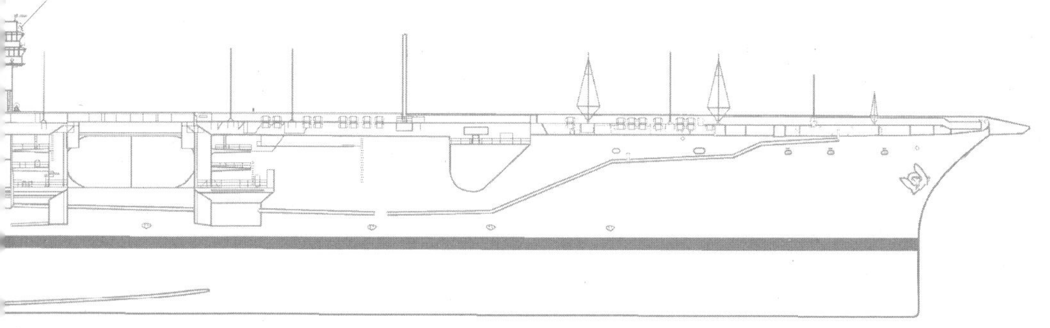

THE SUPERCARRIERS

THE SUPERCARRIERS

The *Forrestal* and *Kitty Hawk* Classes

Andrew Faltum

Naval Institute Press
Annapolis, Maryland

This book has been brought to publication with the generous assistance of Marguerite and Gerry Lenfest.

Naval Institute Press
291 Wood Road
Annapolis, MD 21402

Library of Congress Cataloging-in-Publication Data
Faltum, Andrew,
The supercarriers : the Forrestal and Kitty Hawk classes / Andrew Faltum.
pages cm
Includes bibliographical references and index.
ISBN 978-1-59114-180-8 (hardcover : alk. paper)—ISBN 978-1-61251-770-4 (ebook) 1. Aircraft carriers—United States—History. 2. Kitty Hawk (Aircraft carrier) 3. Forrestal (Aircraft carrier) I. Title.
V874.3.F3526 2014
359.9'435—dc23
2014021816

Unless otherwise noted, all photographs are official United States Navy photographs.

The front end paper depicts the *Saratoga* late in her career as a representative of the *Forrestal* class. The back end paper of the *Constellation* represents the *Kitty Hawk* class. Unless otherwise noted, all illustrations are the works of the author.

♾ Print editions meet the requirements of ANSI/NISO z39.48-1992 (Permanence of Paper).
Printed in the United States of America.

22 21 20 19 18 17 16 15 14 9 8 7 6 5 4 3 2 1
First printing

Contents

Maps

Preface

In 1955, when the USS *Forrestal* was commissioned, it ushered in a new era of naval power as the first of the "super carriers." Throughout the Cold War and into the uncertain and dangerous world that followed, the first question of senior American leadership in response to a crisis has often been, "Where are the carriers?" The *Forrestal* and the carriers that followed her have played a significant role in world affairs for more than fifty years. Although it is customary for the later ships to be treated as separate classes, this book will address all the conventionally powered aircraft carriers as a group and show their evolution through changing operational requirements and developing technology.

As a boy, I was fascinated with naval aviation and aircraft carriers and built many plastic models of various carriers, particularly the *Forrestal* herself. Later in life, on active duty as an air intelligence officer, I served on the *Midway* home ported in Yokosuka, Japan. I subsequently remained in the Naval Reserve Intelligence Program until retiring in 1995. One of my last active duty for training assignments was to spend two weeks on board the *Saratoga* on one of her last at sea periods before her own retirement. Since I was not part of the ship's company or assigned to the air wing, my duties involved an "operational orientation" that a senior commander would not normally receive. It was an indelible experience. It is difficult to explain to anyone who has never gone to sea in a warship, but in writing this book, as in my previous works, I have tried to convey how such a complex thing as an aircraft carrier works and how her crew brings its aircraft, weapons, and components to life. Technical jargon has been avoided where possible, but technical information has been included in the appendices as appropriate. For those with no military experience, an explanation of some of the conventions used in this book is in order. Dates and times are given in military fashion, that is 1 October 1955, rather than October 1, 1955, and 1300 instead of 1:00 p.m. Unless stated otherwise, all dates and times are local and all distances are in nautical miles. This book has been compiled from many sources and though I have tried to resolve any conflicting information wherever possible, any errors in judgment are ultimately mine.

Introduction

The ships of the *Forrestal* class became the first "super carriers" specifically designed to operate jet aircraft to enter service. The basic soundness of design can be seen from the fact that they became the basis for every U.S. carrier that followed. (The ships of the *Kitty Hawk* class were essentially improved *Forrestal* designs.) To understand how the *Forrestal* and subsequent ships came about, it is necessary to first look at the evolution of carrier concepts following World War II. In the aftermath of the greatest conflict in history, the Navy found that it had "worked itself out of a job" in a postwar world where America had a monopoly on atomic weapons and many in the newly created Air Force regarded the other services as anachronisms. Even as the Navy stressed the need for balanced forces and the importance of sea power in the postwar world, it struggled to develop its own nuclear capability. The Navy's first guided missiles were being developed, while naval aviation contended with the introduction of jet aircraft to the fleet and the development of heavy attack aircraft capable of delivering atomic weapons, which proceeded in the face of Air Force opposition.

The Navy had plans for the construction of a new super carrier, the *United States*, that could operate jet aircraft large enough to carry the early atomic bombs, which were nearly as large as the weapons that had been dropped by B-29s on Hiroshima and Nagasaki. Throughout the postwar unification struggle, planning for the *United States* had continued. With the approval of Congress and the White House, funds had been appropriated and the keel of the carrier was laid at Newport News, Virginia, on 18 April 1949. Five days later, while Navy Secretary John L. Sullivan was out of town, Louis A. Johnson, who had replaced James V. Forrestal as Secretary of Defense when he had to resign in March 1949 following a nervous breakdown resulting from overwork, ordered all work on the carrier halted, ostensibly for budgetary reasons. Admiral Louis E. Denfield, the Chief of Naval Operations, learned of the action from a press release and Sullivan resigned in protest. He was replaced as Secretary of the Navy by Francis P. Matthews, who had no experience or appreciation of the Navy, let alone naval aviation. (Matthews was often referred to with scorn as "Rowboat Matthews" by senior naval leaders.) Naval leadership was in a state of shock, and regarded Johnson's actions as part of an Air Force campaign to retain its monopoly on nuclear warfare. These views seemed credible when funds allegedly saved by canceling the *United States* were earmarked to pay for more B-36 bombers. The Navy was repeatedly outvoted in meetings of the Joint Chiefs of Staff and feared that its appropriations would be continually cut. When Captain John G. Crommelin, a distinguished naval pilot, supplied documents to the press showing that key naval officers, including Admiral Denfield, considered Johnson's policies dangerous to national security, the resulting uproar led to a congressional investigation.

In what became known as "The Revolt of the Admirals," naval officers, led by Admiral Arthur W. Radford, attacked Johnson's policies and the Air Force claims for strategic bombing. The Navy especially criticized the B-36 as "a billion dollar blunder." With a range of 5,000 miles, the B-36 was designed to carry an atomic bomb from American bases to any point on the globe without refueling. The B-36 was a piston-engine design that had originated in 1940. Capable of a top speed of only 375 miles an hour and with a service ceiling of 40,000 feet, the B-36 was vulnerable to the new Russian MiG-15 jet fighters then entering service. In October 1949, as the congressional hearings got under way, the Russians exploded an atomic bomb, ending the American nuclear weapon monopoly. In the end, the budget cuts remained, although the B-36 program was trimmed. Admiral Denfield was relieved as Chief of Naval Operations and replaced by Admiral Forrest P. Sherman. Sherman set about tempering many of the hotheads within the Navy and worked to strengthen the Navy's support in Congress and restore funding for naval aviation. Though hard work and perseverance salvaged much of the Navy, Johnson's cuts had left conventional forces, Army, Navy, and tactical air, ill prepared to fight the "limited" war in Korea.

Even while the controversy was being played out, other possibilities presented themselves. The result would be the first of the super carriers, the *Forrestal*.

Super Carrier 1

What would become the *Forrestal* class began as an outgrowth of the canceled *United States* and, even though the purpose had changed from pure nuclear strategic strike by a few large aircraft to a more general purpose design capable of performing tactical missions by a much larger air group of smaller aircraft, they were remarkably similar in appearance. In fact early photographs of models and artist conceptions of the two designs are nearly identical and the initial contract configuration closely resembled that of the earlier ship. The main outward difference was an enclosed "hurricane" bow.[1] The design was to be flush decked with a retractable island, four stacks on the port and four on the starboard side designed to minimize the effects of exhaust gases on flight operations, and four deck-edge elevators: one on the starboard side between the retractable bridge and the stacks, two on the port side, and one at the stern. Four catapults were to be installed: two on the bow and one each in waist positions, port and starboard. Armament included pairs of 5" gun mounts in sponsons at each quarter. As construction of the first ship proceeded, other developments in carrier design, such as the angled deck and steam catapults, were applied while the *Forrestal* was still on the building ways.

The idea of angling the landing area of a carrier flight deck was a simple, but revolutionary one that originated with the British. With the angled deck the traditional way of landing a carrier aircraft, a level approach with power cut to land, could be changed to a power on approach, which allowed pilots to touch down in the arresting gear and immediately apply full power to lift off and go around again if necessary. When the new jet aircraft were introduced after World War II, their jet engines required time to "spool up" to full power. A poor approach often meant hitting the barricades to prevent crashing into aircraft parked forward. During the Korean War the first generation of straight-winged jet aircraft, with their relatively slow approach speeds, could be accommodated with the existing straight deck carriers, but following the Korean War, as the second generation of swept-wing jets entered service, accident rates went up alarmingly. The U.S. Navy first began to give the angled deck serious consideration in 1951. In 1952 the *Midway* and *Wasp* were given superficial modifications to test the concept, and the *Antietam*, an un-modernized *Essex*-class carrier, was fitted with a true angled deck later that year; the first true angled deck landing was accomplished in 1953.[2] As a result of the experience gained, the decision was made to modify the design of the *Forrestal* to accommodate the angled deck.

The hydraulic catapults used in previous carriers were approaching their design limits, and the U.S. Navy was considering alternative technologies to accommodate the growing weight of carrier aircraft. For the most efficient catapult stroke, nearly constant acceleration is desired and, given the length limits involved, the shorter the braking distance, the longer the power stroke can be. While the Americans worked on powder charge designs, the British worked on steam-powered, slotted-cylinder designs. The first full-scale steam catapult was installed on the HMS *Perseus* in 1950. A remarkable feature of this design was a water brake, which could bring a 5,000-pound catapult shuttle to a halt in only five feet.[3]

The third British innovation leading to the success of the *Forrestal* design was the mirror landing system. To take advantage of the capabilities offered by the angled deck and the steam catapult, a new method of controlling aircraft as they came on board had to be developed. A landing signal officer (LSO) could only control one aircraft at a time, and the limitations of the human eye made control using paddles limited to no more than a half mile. The British system used a large mirror, concave about its horizontal axis, positioned alongside the landing area at the edge of the angled flight deck. The mirror pointed astern at the angle of the glide path and was mounted on gimbals connected to the ship's fire control system, which was gyro stabilized. This allowed the mirror to compensate for any motion of the ship. Aft of the mirror a powerful light source was aimed at the mirror so that a cone of light was reflected back along the glide slope. The pilot would see a spot of light, the "ball," when he flew in the middle of the beam. To position his aircraft more precisely, a horizontal row of datum lights was mounted on either side of the mirror. If the pilot was high on the glide path, the ball would appear above the reference lights, if too low, the ball was below the reference lights. Later, the mirror was replaced by a Fresnel lens and colors added to the ball, but the principle of the Optical Landing System (OLS) was the same.[4]

The *United States* had been designed on the basis of having to operate a 100,000-pound jet that would succeed the AJ-1 Savage as a carrier-borne nuclear bomber.[5] (In 1952 the U.S. detonated its first thermonuclear bomb. Shortly after the Korean armistice in 1953, the Russians also exploded what was thought to be a hydrogen bomb. Later, the earlier atomic weapons were included under the term "nuclear weapons" that came into general use.) As newer nuclear weapons were being developed that were smaller in size, the Bureau of Aeronautics selected the 70,000-pound Douglas A3D Skywarrior (later known as the A-3) as its heavy strike bomber, in 1949. With a smaller aircraft, a smaller carrier was possible. Even before the outbreak of the Korean War, Representative Carl Vinson, long a friend of the Navy, informally indicated that Congress might back a smaller carrier. He suggested a size limit of 60,000 tons and, even though no new plans were prepared, the Bureau of Ships (BuShips) continued studies about what design tradeoffs could be made to bring the carrier design under the 60,000-ton limit. These studies formed the basis of what would become the *Forrestal* class when approval for new carriers came through.

In July 1950, following the outbreak of the Korean War, Defense Secretary Johnson offered the Chief of Naval Operations, Admiral Sherman, a new carrier, and in October, Navy Secretary Matthews approved a revised Fiscal Year 1952 (FY52) shipbuilding budget that included the *Forrestal*.[6] The *Forrestal* was initially laid down on 14 July 1952 as CVB-59 (the CVB designation standing for "large aircraft carrier" included the *United States* as CVB-58 and the *Midway*-class carriers), and as the *Forrestal's* keel was being laid, Congress authorized a second large carrier, the *Saratoga*. Another large carrier would be funded each year for the next five years. The *Saratoga* was included in the FY53 shipbuilding program, the *Ranger* in FY54, and the *Independence* in FY55.[7] With the revival in support for aircraft carriers came a redesignation to reflect their mission rather than size. The new ship (along with the *Midway*-class CVBs and the *Essex*-class CV ships and the mothballed *Enterprise*) were reclassified as CVA "attack aircraft carriers" on 1 October 1952.[8] From FY52 onward, construction of a new carrier every year was a major Navy goal. The Joint Chiefs of Staff adopted goal for a 12-carrier force for FY52, which was increased to 14 in 1952. Ultimately a peacetime level of 15 carriers was established.[9]

As the first carrier laid down after World War II to be completed, *Forrestal* had a standard displacement of 60,000 tons, 76,600 tons full load. (Displacement is the actual weight of the ship, since a floating body displaces its own weight in water. Full load displacement includes the weight of the ship with all fuel and stores on board.) With an overall length of 1,039 feet, *Forrestal* was also the largest carrier built up to that time (except for the short lived Japanese *Shinano* of World War II), and was the first to be specifically designed to accommodate jet aircraft.[10] Compared to a modernized *Essex*-class carrier, the *Forrestal* had significantly greater capacities: 70 percent greater ship fuel (2.5 million gallons vs. 1.5 million), 300 percent more aviation fuel (1.3 million gallons vs. 440,000), 154 percent more aviation ordnance (1,650 tons vs. 650) and 15 percent more nuclear weapons storage (150 tons vs. 130). As a result of the *Forrestal's* capabilities, there was a remarkable improvement in the effectiveness of air operations, allowing for rapid aircraft turnaround and increased safety. Studies determined that her size and design allowed her to operate 96 percent of the year compared to 60 percent for an *Essex*-class carrier, and aircraft accident rates were reduced by half.[11]

Propulsion was provided by a 260,000 shaft horsepower (shp) steam-turbine plant with four shafts, four steam turbines, and eight Babcock & Wilcox boilers capable of driving her at 33 knots. The *Forrestal*, as first ship in her class, had a 600 pounds per square inch (psi) plant, but all subsequent ships had 1,200 psi systems that provided 280,000 shp. (The 1,200 psi boiler systems were introduced in 1954 and offered higher efficiency, reduced weight, smaller volume, and simplified maintenance over the 600 psi systems of World War II vintage.)[12]

The *Forrestal*-class carriers were armed with eight Mark 42 5"/54 caliber automatic, dual-purpose (air/surface target) gun mounts, two to a sponson on each quadrant. They were usually controlled remotely from a Mark 68 Gun Fire Control System, or locally from the mount at the One Man Control (OMC) station. (In U.S. naval gun terminology, 5"/54 indicates a gun that fires a projectile five inches in diameter and the barrel is 54 calibers long, i.e., the barrel length is 5" × 54 = 270".) The self-loading gun mounts each weighed about 60 tons, including two drums under the mount holding 40 rounds of semi-fixed case type ammunition (the projectile and the charge are separate). The maximum rate of fire was 40 rounds per minute; the maximum range was about 13 nautical miles, and the maximum altitude was about 50,000 feet. As threats from aircraft and missiles grew, these weapons were less effective and were later removed and replaced in most cases by Mark 29 NATO Sea Sparrow missile launchers and Mark 15 20mm Phalanx Close-In Weapons System (CIWS) gun mounts. The forward sponsons also created slamming effects in rough weather that reduced speed because of the spray. Most of the forward 5" mounts were removed in the 1960s and the sponsons were either removed or redesigned.[13]

Previous American carrier design philosophy called for the hangar deck to be the main strength deck and the flight deck to be superstructure above it. In U.S. naval parlance, the hangar deck was the first deck and the decks immediately below it were the second, third, etc. Above the hangar deck were "levels," the forecastle deck being the "01" level, the gallery deck the "02" level, and the flight deck the "03" level. In both the *Essex* and *Midway* classes this resulted in a hangar deck clearance height of 17'6." The sides of the hangar were kept open for maximum ventilation to allow aircraft to warm up on the hangar deck. In the *Essex* class, the armor protection was provided mainly by the armored hangar deck; in the *Midway* class, the flight deck was also protected by armor. In the *Forrestal* and later classes, the supporting structure of the ship sides went all the way up to the flight deck, which became the main strength deck as well as providing armor protection. The flight deck was now at the "04" level, resulting in a hangar clearance height of 25 feet. Since the sides of the ship hull were part of the load-bearing structure, the large openings in the hull sides for the deck edge elevators had to be carefully designed so as not to weaken the hull.

The hangar itself had two sets of sliding bulkheads that could close off the hangar deck into three bays to contain blast and fires. There were two 25-man air crew ready rooms on the gallery deck to allow air crew to scramble to the forward and waist catapults, a 60-man room in the gallery amidships next to the Combat Information Center (CIC) and four large ready rooms (two 60-man and two 45-man) under the hangar deck with escalators to provide access to the gallery deck.[14]

The change in design to include a large island superstructure solved many problems posed by the flush deck design with its smoke pipes for stack gases and retractable bridge and electronic masts. The electronic suite on the new island included a large SPS-8 height finder radar atop a pedestal on the wheelhouse and a massive pole mast carrying an SPS-12 air search radar with a Tactical Air Navigation (TACAN) beacon at its top. A second large pole mast carried electronic countermeasures (ECM) antennas. These masts were both hinged so that they could be folded down (the larger center mast folded to port and rested on the flight deck while the smaller mast folded aft) for passage under the Brooklyn Bridge, which was a requirement for major naval ships at the time in order to have access to the New York Navy Yard in Brooklyn.[15] An SPN-8 carrier-controlled approach (CCA) radar was mounted on the aft end of the island.[16] (See appendix A for details on radars and electronics.)

Both the *Forrestal* and *Saratoga* were built with two C-7 steam catapults on the bow forward and two C-11 catapults on the port angled deck sponson.[17] The C-7 was a high capacity slotted-cylinder catapult originally designed to use powder charges and was redesigned as a steam catapult based on the success of the British steam catapults. The original version used 600 psi steam because of the limitations of the *Forrestal*'s propulsion plant. Later versions used 1,200 psi steam. The C-11 was the first U.S. steam catapult and was based on the British BXS-1 system, but with higher steam pressure. When the C-11 catapult that was to be on the starboard sponson in the original flush deck design was moved to the port side of the angled deck, it created a problem in that, for structural reasons, the tracks of the two catapults had to be close together. Operationally, this meant that aircraft could be positioned on the waist catapults at the same time, but could not be launched simultaneously.[18] Later ships of the *Forrestal* class, the *Ranger* and *Independence*, were equipped with four C-7 catapults.

The arresting gear on a carrier sets limits on aircraft performance as much as flight deck size and catapult capacity. The *Forrestal*-class carriers were fitted with Mark 7 systems, which were improvements over the World War II vintage Mark 4 and postwar Mark 5 designs and capable of stopping a 50,000-pound aircraft (up to 60,000 pounds in an emergency) at 105 knots (121 mph).[19] When the design was changed from an axial deck to an angled deck this allowed for a reduction in the number of cross deck pendants, which reduced the number of arresting gear engines required, saving both weight and space.[20] Originally there were six pendants, but this was later reduced to four.[21]

There are many stages in the life of a warship from an approved design to a commissioned vessel. In the mid-1950s, when the *Forrestal* and her sisters were built, there were a number of commercial shipyards, as well as Navy Yards, capable of building such major warships as aircraft carriers. Although many components of the ship may have been brought together and assembled beforehand, the laying of the keel is the symbolic formal recognition of the start of a ship's construction. Launching is the point when the ship enters the water for the first time and, by tradition, the ship is christened with the breaking of a bottle of champagne across the bow as the ship slides down the building ways with a splash. About 12 to 18 months before the ship is to be delivered to the Navy, the pre-commissioning crew (sailors who will eventually crew the ship) are selected and ordered to the ship. The balance of the crew typically arrives shortly before delivery. Sea trials are an intense series of tests to show that the performance of the ship meets the Navy's requirements and to demonstrate that all of the equipment installed on board is functioning properly. New construction ships will also undergo builder's trials and acceptance trials prior to delivery, when the official custody of the ship is turned over from the shipyard to the Navy. The commissioning ceremony marks the acceptance of a ship as an operating unit of the Navy, and with the hoisting of the ship's commissioning pennant, the ship comes alive as the crew ceremonially mans the ship. Thereafter the ship is officially referred to as a United States Ship (USS).[22]

The *Forrestal* was ordered from the Newport News Shipbuilding and Drydock Company in Newport News, Virginia, while the *Saratoga* was ordered from the New York Naval Shipyard in New York (commonly referred to locally as the Brooklyn Navy Yard). Apart from the 1,200 psi power plants and some other detail changes, the two ships were very similar in appearance. The *Ranger* and *Independence* that followed were of the same basic design, but among the most noticeable of the changes were their enclosed sterns compared to the "notched" sterns of the first two ships. The *Ranger* had forward gun sponsons that were of a different shape than those on the *Forrestal* and *Saratoga* and she retained these sponsons when her forward 5" guns were removed. She had an all welded aluminum elevator on the port side, unlike the steel structures of the other *Forrestal*-class ships. Also, because the angle of the after end on the flight deck was changed slightly, her overall length increased to 1,046 feet.[23] The *Ranger* was built at Newport News and the *Independence* at the New York Navy Yard. In order to expedite her construction, the *Ranger* was started in a smaller drydock and about four months later her partially completed hull was floated to the larger drydock where the *Forrestal* had been built. The *Independence* began construction in one drydock with her stem toward the head of the dock to allow material to be delivered over a truck ramp from the head of the dock to the hangar deck at the stern. The island and sponsons were not installed to avoid interference with a traveling overhead crane. She was also moved to another drydock for final construction.

The *Forrestal* was launched at Newport News on 11 December 1954, sponsored by Josephine Forrestal, the widow of Defense Secretary Forrestal, and was commissioned on 1 October 1955. Just before her commissioning, the construction cost of the *Forrestal* was estimated to be $218 million. As other ships followed, the growing costs of constructing and operating such large vessels would become the center of debate both within the Navy and the Defense Department. From her home port in Norfolk, the *Forrestal* spent her first year "working up" in intensive training operations off the Virginia Capes and in the Caribbean, often operating out of Mayport, Florida. As the first of her breed, an important part of this process was training aviators to use her advanced facilities. In November 1956 she left Mayport to operate in the eastern Atlantic during the Suez Crisis, ready to enter the Mediterranean if necessary and returned to Norfolk in December.[24] In January 1957 she sailed for her first of many deployments with the Sixth Fleet in the Mediterranean.

The keel of the *Saratoga* was laid down at the Brooklyn Navy Yard on 16 December 1952. She was christened on 8 October 1955, sponsored by Mrs. Charles S. Thomas (a few token feet of water was pumped into the drydock where she was being built so that she could be officially "launched"), and commissioned on 14 April 1956.[25] For the next several months she conducted various engineering, flight, steering, structural, and gunnery tests and in August sailed for the Guantánamo Naval Base in Cuba on her shakedown cruise. In December she returned to New York for yard work and got under way in February for refresher training in the Caribbean before entering her home port in Mayport, Florida. On 6 June 1956 the *Saratoga* hosted President Dwight D. Eisenhower and members of his cabinet to observe operations. For two days, she and eighteen other ships demonstrated air operations, antisubmarine warfare (ASW), guided-missile operations, and the Navy's latest bombing and strafing techniques. A highlight of the visit was the nonstop flight of two F8U Crusaders and two A3D Skywarriors from the *Bon Homme Richard* off the west coast to the *Saratoga* in the Atlantic. (This was the first carrier-to-carrier transcontinental flight and was completed by the F8Us in 3 hours 28 minutes and by the A3Ds in 4 hours 1 minute.)[26] The carrier left Mayport in September 1957 for her maiden transatlantic trip, sailing into the Norwegian Sea to participate in Operation Strikeback, joint naval maneuvers of the North Atlantic Treaty Organization (NATO) countries. She returned briefly to Mayport before entering the Norfolk Naval Shipyard for repairs. In February 1958 *Saratoga* departed Mayport for the Mediterranean and her first deployment with the Sixth Fleet.

Ranger had the distinction of being the first American aircraft carrier to be laid down as an angled-deck ship (*Forrestal* and *Saratoga* had been converted while under construction). She was laid down 2 August 1954 at Newport News; launched 29 September 1956, sponsored by Mrs. Arthur Radford (wife of Admiral Radford, then Chairman of the Joint Chiefs of Staff); and commissioned at the Norfolk Naval Shipyard 10 August 1957. In October 1957 she sailed to Guantánamo, Cuba, for her shakedown, conducting air operations, individual ship exercises, and final acceptance trials along the eastern seaboard and in the Caribbean until June 1958. She then departed Norfolk for a two-month cruise (with 200 Naval Reserve officer candidates on board) that took her around Cape Horn. She arrived at her new home port, Alameda, California, in August and joined the Pacific Fleet. The carrier spent the remainder of 1958 in pilot qualification training and fleet exercises along the California coast. Departing in January 1959 for final training in Hawaiian waters until February, she joined the Seventh Fleet for air operations off Okinawa and maneuvers with Southeast Asian Treaty Organization (SEATO) naval units out of Subic Bay in the Philippines.

The *Independence* was laid down at the New York Naval Yard on 1 July 1955. She was launched on 6 June 1958, sponsored by the wife of Thomas S. Gates, the Secretary of the Navy, and commissioned on 10 January 1959. After conducting shakedown training in the Caribbean, she arrived at her home port, Norfolk, in June 1959. She operated off the Virginia Capes for the next year on training maneuvers, and departed in August 1960 for her first cruise to the Mediterranean.

2 Thunder Ballet

It has been said many times that the most dangerous four and a half acres on earth is the flight deck of an aircraft carrier. This is no exaggeration. Most visitors to a carrier when in port are struck by the vastness of the flight deck. What they do not realize is that before pulling into port, most of the embarked air wing's aircraft have been flown off to operate from airfields ashore.[1] If those same visitors were to observe an operating carrier at sea from an aircraft above, they would as likely be struck by how small the deck seems, especially considering that jet aircraft have to land on it. On the carrier itself, there is an area of the island structure where observers can watch flight operations safely. In the sometimes grim humor of naval aviation, this area is known as "vultures row." What would seem absolute chaos to the uninitiated is actually the carefully orchestrated work of many.

As each of the new super carriers was commissioned, the demanding process of molding the men and machines into an effective weapon began. In naval parlance, this period is called the "shakedown," and in aircraft carriers there is the additional challenge of integrating the air crews and other aviation personnel of the embarked air wing into an effective fighting team. As designed, the *Forrestal* class had a complement of 123 officers and 2,641 enlisted men. The embarked air wing added another 237 officers and 1,675 enlisted men.[2] A large portion of the ship's company on board a carrier are from the "black shoe" Navy, whose officers and men who are trained in the skills of surface warfare. Those from the aviation community are known as "brown shoes." (The term dates from the earliest days of naval aviation when aviators wore brown shoes with their khaki uniforms, while regular line officers would wear blue uniforms with black shoes.)

The captain of a U.S. Navy ship holds absolute responsibility for the combat readiness of his ship and for the safety, well-being, and efficiency of the crew. In practice, the captain delegates the duties of the ship through the executive officer (XO), the department heads, and the officer of the deck (OOD). The XO is next in line for command and, by long tradition and practice, is responsible for matters relating to personnel, routine, and discipline of the ship. Major functions on board are the responsibility of the department heads, with an officer in charge responsible for the readiness of the department. Departments, in turn, are made up of divisions, which may range in size from a few dozen crewmen to perhaps hundreds. In addition to the departments found on every naval warship, such as operations, navigation, weapons, engineering, communications, and supply, aircraft carriers have an air department, headed by the air officer, who is also known as the "air boss." The air boss is responsible, along with his assistant, the "mini-boss," for all aspects of aircraft operations on the hangar deck, the flight deck, and airborne aircraft in the area around the carrier. From Primary Flight Control (PriFly), which is not unlike a tower at an airfield ashore, the air boss and mini-boss maintain visual control of all aircraft operating in the Carrier Control Zone. This zone extends from a five nautical mile radius around the carrier vertically from the surface to infinity and any aircraft operating within the control zone must obtain approval prior to entry.

The flight deck itself is the reason for the very existence of an aircraft carrier and everything that takes place above it or below it in one way or another supports the operation of aircraft. Everyone has a job to do and must do it correctly, every time, in daylight or darkness, cold or heat. Lives and mission accomplishment depend on it. Watching a carrier in operation first-hand is an assault on all the senses. In sunny weather, the sea around, stretching to the horizon, can be so incredibly deep blue that it has to be seen to be believed. At other times it has the glistening gray quality of liquid steel. And at night all is a most profound blackness broken garishly by man-made lights that can lead to disorientation because of the absence of visual terms of reference. Smells fill the nostrils: the greasy, waxy smell of the steam from the catapults, the tang of JP-5 aviation jet fuel (the "JP" is for jet propellant, which is essentially kerosene) and the stack gases.[3] All of these odors assault the newcomer, but those that work on the "roof" become so used to them that they no longer notice. When the carrier turns into the wind to launch or recover aircraft, the pressure of the 30 knots of wind created over the deck can be felt by all hands topside. Everyone must be ever vigilant, keeping their heads "on a swivel," as the saying goes, to avoid being blown into whirling rotors or propellers, or being sucked into a jet intake, or knocked overboard. (The flight deck is about 60 feet above the water; hitting the sea from that height is like hitting concrete if not done properly.) For this reason everyone wears a floatation device, known as a "float coat," while working on the flight deck. But the sound is the most remarkable thing. Everyone wears ear protectors ("Mickey Mouse ears") but the decibel level is so incredible it can vibrate the rib cage like a drum.

Every crew member working on the flight deck wears a colored jersey and helmet ("cranial protector") to identify their assigned role. Yellow jerseys indicate those with authority over the movement of all aircraft on the flight and hangar decks, to include aircraft handling officers, catapult and arresting gear officers, and plane directors. Green indicates catapult and arresting gear crews, air wing maintenance personnel, air wing quality control personnel, cargo handling personnel, Ground Support

Equipment (GSE) troubleshooters, hook runners, and photographers mates. Brown is for air wing plane captains, who are squadron personnel responsible for preparing aircraft for flight. (Plane captains perform many of the functions that would be done by a crew chief in the Air Force.) Blue is for plane handlers, chock walkers, aircraft elevator operators, tractor drivers, messengers, and phone talkers. These aircraft handling crewmen are under the supervision of the "yellow shirts." Chock walkers carry wheel chocks and tie down chains to secure aircraft to the deck. GSE is known as "yellow gear" and includes engine starters, called "huffers," and other support equipment. Red is used for ordnancemen, crash and salvage crews, explosive ordnance disposal (EOD) personnel, and firefighters. Purple is for aviation fuel handlers, known as "grapes." White is for quality assurance (QA) personnel, squadron plane inspectors, landing signal officers (LSOs), air transfer officers (ATOs), liquid oxygen (LOX) crews, safety observers, and medical personnel. Final checkers, who are the last to examine an aircraft before it is launched, wear white jerseys with a black checkerboard pattern.[4]

The catapult officers, also known as "shooters," are naval aviators or naval flight officers responsible for all aspects of catapult maintenance and operation. They ensure that there is sufficient wind over the deck and that the settings for the catapults provide enough steam so that aircraft reach flying speed at the end of the catapult stroke.

The aircraft handling officer, also known as the aircraft handler (or just "the handler" or sometimes "the mangler," particularly if there has been an accident involving the movement of aircraft), is responsible for "spotting" aircraft on the flight and hangar decks. The challenge is to avoid a "locked deck," where too many aircraft are placed so that no more can land without a re-spot. In Flight Deck Control, in the island structure at flight deck level, the handler uses the "Ouija board," a scale diagram of the flight deck with flat cutouts representing the various aircraft on board. Different cutouts can depict aircraft with wings folded or extended. The hangar deck is also represented to help coordinate movement to and from the flight deck via the four deck edge aircraft elevators.

Aircraft directors, who may be commissioned officers or enlisted aviation boatswain's mates, are responsible for directing all aircraft movement on the hangar and flight decks. During flight operations, or when the deck is being re-spotted, there may be a dozen or more yellow shirts on the flight deck, reporting directly to the handler. Directors use a complex set of hand signals (with lighted wands at night) to direct aircraft. Aircraft directors are used at airfields ashore, but the demands of guiding taxiing aircraft, often with only inches of clearance, on a rolling and pitching deck, with the wind, whirling propellers and rotors, and jet engine exhaust, and sometimes in the rain or darkness, is several orders of magnitude greater.

Aircraft carriers normally launch and recover aircraft in groups. Each of these "events" may typically include a dozen or more aircraft and are numbered in sequence throughout the day's flight operations. Before flight operations begin, aircraft on the flight deck are spotted so that the first event aircraft can easily move to the catapults once they have been started and inspected. After the first event is launched, which usually takes about fifteen minutes, the second event aircraft are readied for launch about an hour later. Launching of the second event makes room for the first event aircraft to land. After the first event is recovered the aircraft can be refueled, re-armed, and re-spotted for the third event. Any aircraft requiring maintenance are sent to the hangar deck below. This process is known as "cyclic operations," and the length of each cycle can vary from the normal hour and a half, depending on factors such as the operational tempo and the number of aircraft involved. If the cycle is shorter, fewer aircraft can be launched or recovered; if the cycle is longer, fuel may become more critical for airborne aircraft.

About 45 minutes before launch, each flight crew conducts a walk-around inspection of their aircraft before manning it. About half an hour before launch, aircraft are started and pre-flight inspections are completed. Around 15 minutes later, aircraft are taxied from their parked positions and spotted near the catapults. As the ship turns into the wind, the first aircraft move onto the catapults and extend their folded wings. Behind them, as they move into position on the catapults, Jet Blast Deflector (JBD) panels are raised. Final checkers make a last visual inspection before final catapult hook-up and ordnancemen arm any loaded weapons.

When the *Forrestal* and her sisters were first commissioned, the method of hooking up to the catapult was accomplished by positioning a wire rope cable called a "bridle" onto the catapult shuttle, which was in turn attached to the catapult gear under the flight deck. The two ends of the bridle were then attached to hooks on the aircraft. (In cases where only a single attachment point on the aircraft was used, the cable was known as a "pendant.") The hold-back, a fitting that keeps the aircraft from moving forward prior to catapult firing, is then attached and the catapult is put under tension to take all the slack out of the system. (The bridle system was in use since World War II and the bridles were initially used only once. Later, an elastic strap called the slide lanyard was attached to the bridle so that it could be re-used. If the bridle was damaged or had been used enough it was discarded. On many carriers there were structures called "bridle catchers" attached to the bow to allow the bridles to be recovered for reuse. Once all aircraft on board carriers were equipped to use the nose-tow system, which was first introduced in 1962, bridles were no longer necessary and many carriers have had their bridle catchers removed as they were modernized or overhauled.)[5]

Once the catapult is tensioned, the pilot is signaled to advance the throttles to full military power and he takes his feet off the brakes. After checking his engine instruments and moving all the control surfaces to see that they are operating properly, the pilot signals that his aircraft is ready for flight by saluting the catapult officer. (At night, he turns on the aircraft exterior lights.) Meanwhile final checkers are visually inspecting the aircraft for proper flight control movement, engine response, any leaks, and that access panels are secure. If satisfied, the checkers give a thumbs up to the catapult officer. After a final check of catapult settings, the "shooter" gives the signal to launch and the catapult operator then pushes the firing button. Once the catapult fires, the hold-back breaks free as the shuttle moves rapidly forward, dragging the aircraft with it. The aircraft accelerates from zero to about 150 knots in a few seconds. (This acceleration is relative to the carrier deck; wind over the flight deck gives the aircraft additional lift.)

Just after becoming airborne, the pilot raises the landing gear and makes a ten degree clearing turn, to the right if launched off the bow and to the left if launched off the waist catapults, which increases the separation from other aircraft being launched. After the clearing turn, aircraft turn parallel to the ship's course and proceed straight ahead at 500 feet out to seven nautical miles. Then, depending on visual conditions, aircraft are either cleared to climb to their assigned altitudes or turn to intercept a ten nautical mile arc around the ship and maintain visual conditions until they are established on their outbound departure radial, when they are free to climb up through the weather. While all aircraft of the event are launched, preparations for launching the next event are under way and the process repeats itself throughout the day's operations.

As exciting as being launched from an aircraft carrier is, the most challenging job in naval aviation is landing on board again, especially in foul weather, under poor visibility, or when there are other conditions, such as aircraft malfunctions, to complicate the process. (Medical studies that monitored the physical reactions of pilots to stress have shown that actual combat is less stressful than landing on a carrier at night, for example. For this reason, landing on board at night in bad weather is often referred to as "practicing bleeding.") The *Forrestal* class had been intended from the beginning to be capable of operating as all-weather attack carriers, but when they first joined the fleet, the Navy had neither truly all-weather aircraft nor a shipboard system that would guide them all the way down onto the flight deck in conditions of bad weather and poor visibility.[6] The Navy had been developing doctrine for Carrier Control Approach (CCA) landings, the naval equivalent of Ground Control Approach (GCA) landings at airfields ashore, since 1948. The first production All-Weather Carrier Landing Systems (AWCLS) were installed on the *Midway* and *Independence* in 1962 and approved the next year to be installed on all the Navy aircraft carriers operational at the time. This system was the SPN-10, and evolved over the years through the SPN-42 and subsequent SPN-46 to become the Automatic Carrier Landing System (ACLS) currently in use. (These systems will be covered in more detail in chapter 7.)[7]

Under normal conditions in good weather, as aircraft from returning events wait to be recovered, they enter a left turn "holding pattern" off the ship's port beam, usually by flights of two or more aircraft "stacked" at various altitudes based on their aircraft type and squadron. The minimum holding altitude is 2,000 feet and there is normally a minimum vertical separation of 1,000 feet between flights. Flights arrange themselves for proper separation and, as aircraft from the follow-on event launch and the flight deck is cleared, the flight on the bottom leaves the stack and descends so as to arrive at the "initial" point, three miles astern of the ship at 800 feet and on a course parallel to the ship's. (As the first flight departs the stack, each of the flights above descend to the altitude of the flight below.) From the initial point the flight flies over the ship and each aircraft "breaks" into the landing pattern at about one minute intervals to establish proper separation. (If too many aircraft are in the landing pattern when a flight arrives at the ship, the flight leader initiates a "spin," a tight, slightly climbing, 360 degree turn within three miles of the ship. The break is a level 180 degree turn made at 800 feet. Once established downwind the pilot descends to 600 feet, lowers the landing gear and flaps, and completes the landing checks. When abeam of the ship, the aircraft is about a mile and a half away. This point is known as the "180" (actually closer to 190 degrees because of the angle of the flight deck). The pilot begins his turn to final while beginning a gentle descent. At the "90" the aircraft is at 450 feet, about a mile from the ship. As the pilot crosses the ship's wake, the aircraft should be approaching final and at about 350 feet. At this point, the pilot "calls the ball" of the Optical Landing System (OLS) and devotes his attention to maintaining proper glide slope, lineup, and angle of attack until touchdown. During aircraft recovery, the arresting gear engines must be set for varying resistance based on the type and weight of each aircraft landing. The arresting gear officer (AGO), who is responsible for monitoring operation of the arresting gear and the status of the landing area, ensures that the deck is ready. In naval parlance, the flight deck is either "clear" and ready to land aircraft or "foul" and not ready for landing.

Aircraft on final approach are under the control of the landing signal officer (LSO), who is responsible for ensuring that approaching aircraft are properly configured, and on the correct glide path angle, attitude, and lineup. LSOs, qualified, experienced pilots themselves, communicate with landing pilots via voice radio and light signals. The OLS has green cut lights used during a "zip-lip" (no radio communications) approach flashed for a few seconds to indicate that the aircraft is cleared to continue the approach. Subsequent flashes of the cut lights prompt the pilot to add power—the longer the lights are left on, the more power should be added. Red wave-off lights indicate that the pilot must add full power and go around—a mandatory command. Since the landing area is only about 120 feet wide and aircraft are often parked within a few feet on either side, lining up on the centerline is critical.

Immediately upon touchdown, the pilot advances the throttles to full power so that, if the hook fails to catch a wire (cross deck pendant) a go-around, known as a "bolter," can be executed. If everything goes right, the tailhook catches the target wire (the ideal landing is for the hook to engage the third of the four wires, which abruptly brings the aircraft to a full stop in about two seconds. As the aircraft's forward motion stops, the engines are throttled back to idle, and the hook is raised on the aircraft director's signal, after which the pilot is directed to taxi clear of the landing area to prepare for the next aircraft. The hook runners ensure that the wire is clear of the aircraft's tailhook. (The cross deck pendants are replaced after so many landings and are routinely inspected for unusual wear. A wire that snaps during an arrested landing whips about with such force that it can cut a man in two.) The wings are folded and the aircraft is taxied to a parking spot and shut down, after which it may be refueled, re-armed, and inspected. Any remaining ordnance will be de-armed, minor maintenance may be performed, and the aircraft is often re-spotted prior to the next launch cycle.

During flight operations a ship, usually a destroyer, is assigned as a plane guard to recover the crew of planes that crash or ditch in the water. The plane guard is normally stationed at least 3,000 yards behind the carrier and either to port and clear of the carrier, or at some point that intersects the carrier's final approach line. This can be dangerous when either ship does not properly anticipate the moves of the other, especially since carriers must often change speed and course to stay headed into the wind. A plane guard can find itself under the bows of a carrier traveling at full speed. Before the Korean War began, the Navy discovered that helicopters were more efficient and effective as plane guards, since they could get there faster and safer, but night operations still required a ship as plane guard.

Flight operations in a carrier are only part of the bigger picture. There is a continual process of preparation, execution, and follow-up. The usual cycle takes about 18 months—six months working up, six months deployed to a forward area, and six months preparing for the next deployment. The cycle normally begins when the ship is returning to its home port from overseas deployment. After a leave and upkeep period, followed by local at-sea operations, the ship undergoes a planned maintenance "availability" during which most repairs and equipment upgrades are completed.[8] When ready for sea, the carrier begins working up for its next deployment by completing training exercises and evolutions that become more complex as the crew's proficiency increases. When the carrier and the embarked air wing are ready, the ship begins battle group training with the staffs and other units in the carrier battle group, while continuing proficiency training for the ship's company and air wing. Of course, the cycle can be shortened and ships deployed sooner than planned when a crisis develops somewhere in the world and the national leadership decides to bring additional naval forces to bear. During deployment, a carrier battle group usually includes the carrier, two guided-missile cruisers, two guided-missile destroyers, and one or two ASW destroyers or frigates. (During much of the 1990s and up to 2003, a carrier group was usually sent out with an Amphibious Ready Group [ARG]. The ARG would have a big deck Amphibious Assault Ship [LHA] or Landing Helicopter Dock [LHD] ship and three supporting amphibious ships with about 2,000 Marines of a Marine Expeditionary Unit embarked.) In 2003 the term Carrier Battle Group (CVBG) was changed to Carrier Strike Group (CSG).

The Golden Age 3

Aircraft carriers are unique among warships in that the aircraft they carry are their "main battery." Since the early days of naval aviation, advances in carrier aircraft design have, in turn, had an effect on the operational effectiveness of the carriers themselves. Also, as these new operational capabilities are developed, tactics, techniques, doctrine, and organizations have evolved to take full advantage of them. The basic unit of naval aviation is the squadron. During the time the *Forrestal* and her sisters first joined the fleet these included fighter (VF) and attack (VA) squadrons that, along with detachments of specialty aircraft (known as "overhead") made up the Carrier Air Groups (CVGs). In 1963 the air groups of attack carriers were redesignated as Carrier Air Wings (CVWs) and have evolved as newer and better carrier aircraft appeared and new weapons and capabilities were developed. (See appendix D.)

Fighter Aircraft

The straight-wing F2H Banshees and F9F Panthers that filled most carrier fighter squadrons at the end of the Korean War were replaced with the second generation of swept-wing fighters. Swept wings allowed jets to deal with the Mach effects of supersonic flight and the performance of carrier-based aircraft once again compared favorably with their land-based counterparts. These second generation fighters included the nuclear-capable F9F-6 Cougar, the F7U-3 Cutlass, and later the FJ-4B Fury. (Initially, Composite Squadron [VC] detachments of F2H-4 Banshees provided night fighter capabilities to the new carriers until the development of true all-weather fighters.)

The Grumman F9F-6 Cougar model of the single-engine F9F Panther introduced swept-wing fighters to the carrier air groups. The Cougar series reached the F9F-8 version and was used primarily as a day fighter, but there were light attack and photo reconnaissance versions as well. Although the Cougar was replaced as a fighter on the carriers by the F11F and F8U in the late 1950s, training versions remained in naval service until 1974.

Among the first post–Korean War fighter deliveries were improved, swept-wing FJ-2 models of the North American Fury. Ironically, this was a "navalized" version of the Air Force's F-86E, while the original XF-86 prototype design evolved from the Navy's straight-winged FJ-1. The FJ-3 variant of the Fury appeared in time to be the first aircraft to operate from the *Forrestal*. It featured a new wing, fuselage, rudder, engine, and landing gear. With six underwing carrying points, the FJ-3 was an excellent weapons platform for the Sidewinder air-to-air missile. The FJ-4 Fury was a complete redesign with new fuselage, wing, and control surfaces. With six wing store points, additional armor, a low-altitude bombing system, and an improved control system for high speed flight at low altitude the FJ-4B Fury was a mainstay of many carrier attack squadrons and was the only version capable of delivering nuclear weapons.

Three or four years behind the second-generation jet fighters came several new designs in the mid- to late 1950s into the early 1960s. These included the supersonic Vought F7U Cutlass, the Douglas F4D Skyray and Grumman F11F Tiger, and the McDonnell F3H Demon. A trailing member of this generation was the Vought F8U Crusader, which would remain in front line service into the 1970s, with reconnaissance versions serving into the 1980s.

The Vought F7U Cutlass was one of the most unorthodox aircraft ever to operate from a carrier. It was a tail-less, single-seat, twin-engine jet with twin vertical tail fins but no horizontal tail surfaces. The first versions of the Cutlass were underpowered (leading to the nickname "Gutless Cutlass") and the introduction of improved engines was delayed. Although the F7U-3 version could reach speeds exceeding Mach 1, the speed of sound, the radical design encountered previously unknown aerodynamic problems, which earned it the nickname of "The Ensign killer." Few squadrons made deployments with the Cutlass, and most beached them ashore during part of the cruise because of operational problems, but the VA-86 Sidewinders did deploy on the *Forrestal* from January to March 1956 during her shakedown cruise. The F7U-3 version could carry a nuclear weapon and F7U-3P photo reconnaissance and F7U-3M missile-armed variants followed. (The F7U-3M and F3H-2M were the first to use the Sparrow air-to-air missile on board carriers.)[1]

During this period, the Navy also introduced two new all-weather fighters, the Douglas F4D Skyray and McDonnell F3H Demon, into the fleet. As the Navy's first delta wing fighter, the F4D was a single-seat, single-engine high-performance interceptor that sacrificed endurance for rate of climb and speed in order to counter the threat of high-altitude bomber attacks on carriers. After engine changes and modifications the "Ford," as it soon became known, began capturing flight records. In addition to serving in Navy and Marine all-weather fighter squadrons, the F4D also served with the shore-based Navy all-weather fighter squadron VFAW-3 as part of the North American Air Defense Command (NORAD), a predominantly Air Force organization.

The McDonnell F3H Demon was a swept-wing, single-engine, single-seat fighter that was developed in parallel with the Skyray. The Demon first flew in 1951 and early models were underpowered and accident prone. After production of the improved F3H-2N Demon began in 1955, the fighter found its place in carrier air wings. In 1956 VF-14 received the F3H-2N and the Tophatters flew them from the *Forrestal* during a January to July 1957 Mediterranean deployment. The Navy eventually equipped 22 squadrons and the Demon also deployed on board *Saratoga*, *Ranger*, *Independence*, and *Constellation* until 1963. The F3H-2M missile-armed version was equipped with Sparrow air-to-air missiles. Four Sparrows could be carried, but a full load was rarely carried by fleet aircraft because their weight seriously affected performance. Although not supersonic, the Demon complemented day fighters such as the Vought F8U Crusader and Grumman F11F Tiger as an all-weather, missile-armed interceptor.

The F11F Tiger was an outgrowth of Grumman efforts to incorporate new aerodynamic concepts, such as area rule, and other advances to the Cougar design. In the end, the Tiger was a completely new aircraft. Although supersonic, the Tiger's carrier service lasted only four years because of its engine reliability problems and lack of range and endurance. Tigers served in seven Navy fighter squadrons and operated from the *Forrestal*, *Saratoga*, and *Ranger*. Its place was taken by the superior Crusader, which entered fleet service at about the same time in 1957.[2]

After the Navy's bad experiences with the Cutlass, Vought came up with a winner in the F8U Crusader. The design had several innovative features, but the most notable was the variable incidence wing. (Not to be confused with the *variable sweep* wings of later aircraft, such as the F-14 Tomcat.) In order to increase the angle of attack for greater lift, the wing pivoted seven degrees out of the fuselage. This kept the fuselage level to maintain cockpit visibility during carrier launches and recoveries. Further lift was provided by drooping leading-edge slats and extending the inboard flaps. The Crusader was the last Navy fighter designed with cannons as its primary armament, although it had launch rails on "cheek" pylons for Sidewinder air-to-air missiles.[3] The aircraft exceeded the speed of sound during its maiden flight and the development was so trouble-free that the second prototype and the first production F8U-1 flew on the same day. In April 1956 the F8U-1 performed its first catapult launch from the *Forrestal*. The Swordsmen of VF-32 were the first fleet squadron to fly the Crusader and deployed to the Mediterranean in late 1957 on the *Saratoga*. Navy fighter pilots were happy with the Crusader—it was a fighter. It could reach Mach 1.7, had a phenomenal rate of climb, and was maneuverable. But the Crusader was not an easy aircraft to fly. Although it took a catapult shot well, it was often unforgiving in carrier landings, particularly on the smaller *Essex*-class carriers, because of yaw instability. Later versions had twin ventral strakes fitted under the tail to increase directional stability, but the improvement was negligible. Once on board, the castered nose wheel caused problems steering on the deck. Moreover, with its short landing gear, which caused the exhaust to be close to the deck, and low mounted air scoop that looked like jaws ready to suck up the careless, it was called the "Gator" by wary flight deck crews. While known as the ultimate "day fighter" later versions of the Crusader had limited all-weather and strike capabilities. As the Crusader was replaced by the Phantom II on the larger ships, it remained in service on board the modernized *Essex*-class ships still operating as attack carriers until 1976. The reconnaissance versions of the Crusader continued in service until 1987.[4]

Aerial Refueling

In 1955 the Navy decided that all fighters in production would be fitted for in-flight refueling and many aircraft were modified accordingly. This technique extended the range of strike aircraft, allowed combat air patrols to stay aloft longer, and enabled aircraft returning from a mission to be refueled while flight decks were cleared or other aircraft landed or launched. The tanker aircraft would unreel a hose with a flexible cone, called a "drogue," at the end. The pilot of the receiving aircraft would bring his plane below and aft of the tanker, adjust his speed to that of the tanker, and jockey his plane's refueling probe into the trailing drogue. The probe and drogue would automatically interlock and the fuel would be transferred. (The Air Force developed another aerial refueling method using a flying boom, which offered faster fuel transfer, but required a dedicated aircraft with a boom operator station. The Navy method was simpler to adapt to existing aircraft, which could use refueling pods and additional drop tanks on other aircraft.) Initially, the refueling probes were mounted externally on existing aircraft, but later aircraft were equipped with retractable probes.[5]

Ejection Seats

In the early days of jet aircraft development, many aircraft manufacturers had designed their own ejection seat escape systems, each with varying characteristics and performance. However, following the successful demonstration of a ground level ejection seat designed by the British Martin-Baker company at the Naval Air Test Facility in Patuxent River, Maryland, the Navy decided to standardize on the Martin-Baker Mark 5 for all its jet fighters and trainers. Subsequently, most Navy tactical jet aircraft were equipped with Martin-Baker "zero-zero" (zero altitude and zero airspeed) ejection seats.[6]

Attack Aircraft

After World War II the Navy combined its separate torpedo and bomber squadrons into a new category of "attack" aircraft. Although it entered service too late to see combat in that war, the Douglas AD Skyraider, a propeller-driven "dump truck with wings," served admirably in the Korean War and was known as the "Able Dog" (a play on the phonetic code for AD). Redesignated as the A-1, the "Spad," as it later became known, soldiered on into the 1960s on the large carriers and continued to operate from the smaller *Essex*-class carriers after that. The Skyraider also served with many other air forces and special variants were developed for electronic warfare, airborne early warning, and night attack.

The Douglas A3D Skywarrior was the result of the Navy's efforts to find a pure jet successor to the North American piston-jet AJ Savage. The Navy initially assumed that an aircraft capable of launching from a large carrier to deliver a nuclear weapon over distances and at speeds comparable to land-based jet bombers would have a gross weight of about 100,000 pounds. When Ed Heineman, chief engineer at Douglas, came up with a design that was under 70,000 pounds, the Navy was at first skeptical, but the Skywarrior went on to become a most successful aircraft, the largest carrier aircraft developed at the time. The Skywarrior had shoulder-mounted wings swept at 36 degrees with the engines slung beneath them in nacelles. The wings folded outboard of the jet engines and the tail folded down to starboard. (As part of the weight-saving measures, the three man crew did not have ejection seats, but bailed out by sliding down a hatch in the bottom of the fuselage.) The prototype XA3D-1 first flew in October 1952 and the first squadrons were equipped in 1956. Heavy Attack Squadrons (VAHs) of seven to ten aircraft operated from the *Forrestal*- and *Midway*-class carriers to give the fleet an all-weather nuclear strike capability.[7] Electronic warfare and photo versions were also developed. As the Navy developed its Polaris-equipped ballistic missile submarines, the A-3, as it was redesignated in 1962, transitioned to other roles, primarily as electronic warfare and tanker aircraft. Because of its size, the A-3 was known as the "whale." The last EA-3B electronic warfare versions served until 1991.[7]

Before the Skywarrior, the Navy's nuclear heavy attack capability was provided by the North American AJ Savage, an aircraft powered by two piston engines for normal operations with a jet engine for bursts of speed during combat.[8] In practice, the Savages, which entered service in 1951, did not operate on board the carriers for extended periods, but were kept ready at advanced land bases and flown out to the carriers if the situation required. As the Skywarrior joined the fleet it replaced the Savage in the heavy attack role, but a few served into the early 1960s as aerial tankers and photo planes.[9]

The Douglas A4D Skyhawk and the A3D Skywarrior were the Navy's first jet attack aircraft and were literally at opposite ends of the scale, even though both were developed by the same designer, Ed Heineman. In an era when carrier aircraft were getting bigger and heavier, the Skyhawk, remarkable for its small size, was known as the "scooter" or "Heineman's Hot Rod." (It was small enough that it did not need to have folding wings.) The A-4, as it was later designated, was a single engine delta wing jet with tall tricycle landing gear that allowed it to carry a wide variety of ordnance, including the new, smaller nuclear weapons.[10] Through various improved versions, and adaptations to other roles, the Skyhawk was in production until 1979 and went on to serve with many air forces. TA-4J training versions of the Skyhawk remained in Navy service as target-towing, adversary, and combat training aircraft until 2003 and the Skyhawk continues to serve in some foreign air forces.[11]

Airborne Early Warning

The Navy's experience with the Japanese kamikaze attacks in World War II and the concern over the increased threat posed to carrier task forces by jet aircraft and missiles in the postwar era, led to the development of carrier-based airborne early warning aircraft, starting with the TBM Avenger during World War II, through the Grumman Guardian and radar-equipped variants of the Skyraider in the postwar era.[12] These were modifications of existing single engine carrier aircraft and provided only a capability for early warning. With the continuous improvements in early airborne radars, the Navy decided in 1956 to develop an airborne early warning and command and control aircraft, the Grumman WF Tracer. The WF, as the Navy's first purpose-built carrier airborne early warning aircraft, entered service in 1958. The Tracer was a derivative of the Grumman TF Trader, which in turn was derived from the Grumman S2F Tracker, a twin piston engine carrier-based antisubmarine aircraft. Major changes included replacing the single large vertical tail with a twin tail and folding the wings back along the fuselage in typical Grumman style (the wings of both the Tracker and Trader folded upwards over the fuselage), but the most distinctive feature was the large aerodynamically shaped radome that housed the APS-82 radar. This radar had many new features for the time, such as a stabilized antenna and an Airborne Moving Target Indicator (AMTI) capability that allowed the radar to detect low-flying targets against the clutter of radar reflections from the ocean. The Tracker was initially designated the S2F (later S-2) and was known as the "Stoof." The Tracer was initially designated the WF (later E-1) so it became known as the "Willy Fudd" or, because of its large antenna radome above the fuselage, the "Stoof With a Roof." The Tracer began to be replaced by the more modern Grumman E-2 Hawkeye in the early 1970s, but continued to serve on carriers until 1977.[13] (The E-2 will be discussed in chapter 6.)

Helicopters

At the time the first ships of the *Forrestal* class were entering service, the standard shipboard helicopter for plane guard and utility work on carriers was the Piasecki HUP Retriever, a compact single-engine, twin-tandem-rotor utility helicopter that was introduced in the early 1950s and served until 1964.[14]

In the late 1950s the Navy awarded Kaman Aircraft Corporation a contract to develop a fast, all-weather utility helicopter. The Kaman HU2K-1 design featured four blades on the main rotor and three blades on the tail rotor with a single turboshaft engine. By the time it entered service in late 1962 it was designated the UH-2 Seasprite and was primarily deployed on board carriers in search-and-rescue (SAR) and plane guard roles. (The UH-2 was selected to be the basis for the interim Light Airborne Multi-Purpose System [LAMPS] helicopter in October 1970 and later versions served until 1993.)

Introduced in 1961 as the HSS, the Sikorsky Sea King was one of the most successful helicopter designs in history. It was the first ASW helicopter to use turboshaft engines, as well as the world's first amphibious helicopter. Sikorsky built over a thousand and it was produced under license by Italy, Japan, and the United Kingdom. The SH-3, as it was later known, has been adapted for many other roles, such as search and rescue, transport, anti-shipping, and airborne early warning as well as plane guard. The last SH-3 was retired from U.S. service in 2006, but continues to serve with other countries.

The Boeing Vertol H-46 Sea Knight twin-tandem-rotor helicopter was introduced in 1964 as a medium-lift helicopter to replace the earlier single-rotor Sikorsky designs for the Marines. Powered by twin turboshaft engines, it was capable of carrying up to 25 troops and had a hook beneath the fuselage capable of carrying up to 10,000 pounds as a sling load. Experimentation by the Navy in vertical replenishment (VERTREP) of ships under way at sea with single rotor helicopters had proven disappointing, but the twin-tandem design of the H-46 showed that it could operate in a wider range of wind conditions. Known as the "Phrog" by the Marines, it was extremely maneuverable and responsive and was selected by the Navy in its UH-46 configuration to operate from replenishment ships supporting the carriers. The first VERTREPs were conducted in the Mediterranean in 1965 and have become standard operations since then. While the Navy retired the UH-46 in 2004, replacing it with the H-60, the Marines will continue to operate them until the MV-22 Osprey is fully fielded.[16] (The H-60 is covered in chapter 11.)

Sparrow

The original Sparrow evolved from a Navy program of the late 1940s to develop an air-to-air guided rocket. After protracted development, the first AAM-N-2 Sparrows entered limited operational service in 1954 and, in 1956, were carried by the F3H-2M Demon and F7U Cutlass. The Sparrow I was a limited and crude weapon. Its beam-riding guidance restricted the missile to attacks against targets flying a straight course and it was useless against a maneuvering target. The AAM-N-6 Sparrow III, developed concurrently with the Sparrow I, was a semi-active radar homing version that entered Navy service in 1958 and became the basis for later missiles.[15] In 1963 the Navy and Air Force agreed on a common naming system for their missiles and the Sparrow became the AIM-7.[17]

Sidewinder

The Sidewinder was the result of an in-house effort on the part of the Naval Ordnance Test Station (now the Naval Air Weapons Station China Lake) to develop a heat-seeking missile with the "electronic complexity of a table model radio and the mechanical complexity of a washing machine."[18] The result was a reliable and effective missile that has improved over time and has become one of the most widely used air-to-air missiles in the world. Originally designated the AAM-N-7, the Sidewinder went operational in 1956. (The missile was named for a variety of rattlesnake common in the American southwest that hunts by sensing its warm blooded prey. The corkscrew path of the early missiles as they made their course corrections also resembled the way the snake moves in the sand.) Redesignated as the AIM-9, the Air Force also developed their own versions of the Sidewinder, and, over time, the missile has been improved for enhanced performance in dog fights, resistance to countermeasures, and increased lethality.

Bullpup

As a result of its experiences trying to destroy heavily defended targets during the Korean War, the Navy developed the first mass produced air-to-surface command guided missile, the ASM-N-7 Bullpup (later redesignated as the AGM-12B). Using a flare attached to the rear of the missile for visual tracking, the pilot or operator sent commands by a control joystick to guide the missile to the target. The Bullpup first deployed with the Seventh Fleet in the Western Pacific in 1959 with the Rampant Raiders of VA-212, an FJ-4B Fury squadron on the *Lexington*. The following August, VA-34, the Blue Blasters, an A4D squadron on the *Saratoga* joined the Sixth Fleet in the Mediterranean.[19] Unfortunately, the Bullpup worked best when the delivering aircraft and the missile were on the same track. Once in combat use in Vietnam, however, enemy gunners quickly learned that firing into the path of the flare's smoke trail would often hit the delivering aircraft. Another problem was that with only a 250-pound warhead, the Bullpup could not take out hard targets, like concrete and steel bridges. Later versions had improved motors and guidance with a larger 1,000-pound warhead. (There was even a nuclear warhead version.) The Bullpup was phased out of U.S. service in the 1970s but continued to be used by foreign air forces for some years later.[20]

Zuni

The Zuni rocket was another successful product of the Navy's China Lake team. Originally developed for both air-to-air and air-to-ground applications, the Zuni is a 5" unguided folding fin rocket that is launched from the LAU-10 launcher, an aluminum cylinder with frangible end caps that carries four launch tubes. The Zuni has been produced since 1957 and replaced the "Holy Moses" 5" High Velocity Aircraft Rocket (HVAR). (The HVAR was also developed at China Lake during World War II and remained in service into the 1960s.) Although accidents involving the Zuni have played a significant role in a number of carrier fires, the Zuni has been successively improved over time and is still in use.

Walleye

The AGM-62 Walleye was a television-guided glide bomb used in the 1960s. Most had a 250-pound high-explosive warhead, some had a nuclear warhead, and there was a 1,000-pound version. Even though it was officially an "air-to-ground missile" that was misleading, since it was an unpowered bomb with guidance avionics. It was used with some early success in Vietnam in 1967 against softer targets, but sturdier targets such as railroad bridges could not be brought down even with the 1,000-pound version. Accordingly, China Lake developed the 2,000-pound Walleye II "Fat Albert" version, which was used during the 1972 Linebacker strikes against Hanoi and Haiphong. (The Walleye was superseded by the AGM-65 Maverick).[21]

Rockeye II

The Mark 20 Rockeye II is a free-fall, unguided cluster bomb designed to kill tanks and armored vehicles developed by China Lake and fielded in 1968. It uses 247 Mark 118 dual-purpose armor-piercing shaped-charge bomblets, each of which weighs 1.32 pounds, in a Mark 7 dispenser. Rockeye is most effective against area targets requiring penetration to kill. Fielded in 1968, Rockeyes were not widely used until Desert Storm when the Marines used them extensively against armor, artillery, and antipersonnel targets. (The rest were dropped by the Air Force and Navy.)[22]

4 The *Kitty Hawk*

When Dwight D. Eisenhower came into office in January 1953, his concern for balancing the Cold War military commitments of the United States with the nation's financial resources led to an approach to national security that emphasized reliance on "massive retaliation," principally by strategic bombers, to deter potential threats, both conventional and nuclear, from the Soviet Union and its allies. This "New Look" approach was an attempt to take a long-term view of national security and get the "biggest bang for the buck." It would place less emphasis on conventional military forces and result in the Air Force getting almost half the defense budget, while the Navy would get less than a third. (Since budget projections are prepared years in advance, the first budget planning done by the Eisenhower team was for FY55.) During the mid-1950s advances in nuclear warhead design and missile systems led to the development of nuclear-armed intercontinental ballistic missiles (ICBM) and when the Soviets launched their Sputnik satellite in 1957, America became concerned about the perceived "missile gap." Each of the armed services had its own missile programs and, starting in the mid-1950s under the leadership of Admiral Arleigh Burke as Chief of Naval Operations, the Navy began a program to marry the nuclear-powered submarine to the ballistic missile, which eventually became the Polaris.[1] This remarkable development effort was largely "taken out of hide" and Polaris first went operational in 1961. Polaris went on to become one of the legs of the "strategic triad"—land-based bombers, intercontinental missiles, and submarine-launched ballistic missiles.

Admiral Burke, however, was careful to ensure that the Navy's operational flexibility, in the form of aircraft carriers, was not undercut by the success of Polaris. In the 1950s many saw the role of aircraft carriers as primarily one of "power projection" against land targets while critics often regarded their most important function to be one of "sea control" to ensure America could respond to any Soviet aggression in Europe. To these critics, the principal threat posed by the Soviets, who had always had a large submarine force and a strong land-based naval aviation capability, was the threat to convoys of material and manpower to reinforce Europe. President Eisenhower, however, regarded Europe as essentially a large peninsula and saw naval forces as being useful to attack the periphery of Soviet-controlled territory. This would cause the Soviets to disperse their forces over a wider area in order to protect their flanks. (During the Eisenhower administration, nuclear weapons were regarded as just another weapons system and were contemplated for both strategic and tactical roles.) The Soviet response was to develop fast nuclear-powered attack submarines as well as surface- and air-launched anti-ship missiles. Eventually, the rigid massive retaliation approach of the Eisenhower administration would be moderated over time to what would become, under the Kennedy administration, one of "flexible response." Nuclear attack missions for the carrier force would continue (the A3J Vigilante would become the last carrier-based "strategic" attack aircraft, and the only supersonic one), but would eventually diminish in relative importance over the years. In 1960 the Single Integrated Operational Plan (SIOP) was created to integrate the nuclear warfare plans for all the services. With the growing importance of the Polaris and later Poseidon missiles, the carriers were eventually removed from the SIOP altogether in 1976. Nuclear weapons were still carried on board, but now the carriers were part of the nuclear "reserve force" and no longer required to be within distances from launch points and air crews were no longer tied to attacking specific targets in Russia, China, and other communist countries.[2]

As the Cold War developed, the Navy saw in the experiences of World War II several new ominous developments for the future of the aircraft carrier–jet aircraft, guided missiles, and nuclear weapons. As the Soviets developed their capabilities, the Navy responded with the development of guided missiles of its own for defense, both air-to-air and surface-to-air. The shipboard surface-to-air missiles developed into a family of missiles for short, medium, and long ranges. These were the "Three Ts"—the Tartar, Terrier, and Talos. But new weapons were only part of the problem. The Navy's experiences with the saturation kamikaze raids late in the war caused great concern. As the human "guided missiles" of the kamikaze threat were replaced by ever faster and increasingly sophisticated aircraft and guided missiles, the ability of human operators within the Combat Information Centers (CICs) on board ship to process and respond to the threat greatly diminished. The ultimate answer would become the Naval Tactical Data System (NTDS).

The Naval Tactical Data System

Although it was kept classified at the time, task force air defense exercises of the early 1950s had revealed that the U.S. fleet could not cope with the expected Soviet massed air attacks using new high speed jet aircraft and standoff missiles. The Navy's senior leaders, including Secretary of the Navy John Connally (who had been a World War II task force fighter direction officer) and the Chief of Naval Operations (CNO) Admiral Burke, were determined to find a solution. In the mid-1950s every Navy ship had a CIC team that tied task force air defense information together. Radar blips of attacking and friendly aircraft were manually picked off radar scopes and manually plotted on backlit plotting tables; the course and speed of a target was manually calculated from the plots of successive radar blips and then written in by hand near the target's plotted track. If an air target's altitude had been measured by height finding radar, or estimated by "fade zone" techniques, the altitude was also penciled in near the track line.[3] Information on known hostile, known friendly, or unknown tracks, as well as the assigned track numbers, was also penciled in. During World War II, a massed air attack might involve a few hundred raids coming at the task force from all points of the compass and at altitudes from sea level to 35,000 feet. Each ship in the task force had measured and plotted the air targets in their assigned pie-shaped wedge on their radar scopes, and their fighter director officers and gunnery coordinators controlled the fighter-interceptors and ship's anti-aircraft guns within that assigned wedge. Each ship in the task force would also report its track information over voice radio to the task force fighter direction officer. The problem was that all the "moving parts" of the system depended on human interaction and in the new environment the radar plotting teams, the fighter directors, and the gunnery and missile coordinators needed some kind of automated system to process air defense data in an integrated, real-time fashion. In building the NTDS, the Navy faced many formidable technical challenges requiring pioneering efforts in digital computers, transistors, and large scale computer programming, as well as overcoming determined resistance by many senior naval officers.[4] The NTDS system was designed to have components that could fit through the doors and hatches on ship, could be maintained in a harsh operating environment, and could be reprogrammed as necessary. (To gain some perspective as to what was accomplished, The Semi-Automatic Ground Environment [SAGE] system that the Air Force developed to solve similar air defense problems for North America was installed in 40,000 square foot buildings and required 3 million watts of electrical power.)[5] The conceptual framework for NTDS was as daunting as the technical hurdles. As laid out by the development team, NTDS would not rely on special-purpose computers because they would need to be physically reconfigured to change their programming instead of just loading a new program into random access memory. The system would also have to display a number of different symbols to represent friendly, hostile, or unidentified targets, whether it was an air, surface, or submarine target, and many other attributes, such as speed and heading vectors, weapons information, and other graphics. NTDS also had to interact with existing ship and weapon systems and perform analog-to-digital conversions for input and the reverse for output to these systems. A primary data link would connect all the NTDS ships within the task force, but since the system would not be installed on anything smaller than a light cruiser–sized ship, other links would connect the smaller destroyer-sized ships to the net. As air defense was one of the primary reasons for creating the NTDS, an automatic interceptor control data link was included to allow NTDS-equipped ships to automatically steer interceptors to their targets. (The first interceptor to be equipped with what was later known as Link 11 was the McDonnell F4H Phantom II fighter.) Since the Navy already had airborne early warning (AEW) aircraft that carried powerful surveillance radars, it was proposed that analog-to-digital data terminals be developed that convert the radar picture to digital format to be transmitted over the primary data link and, alternately, could transmit the radar picture in analog form to shipboard radar scopes via special shipboard receivers. (The Navy would later go beyond this capability in developing airborne tactical data systems. The system would also be capable of interfacing with systems developed for the Marines.) In March 1961 the *Oriskany*, an *Essex*-class attack carrier with a modular CIC that could accommodate the needed equipment, entered San Francisco Naval Shipyard to begin a five-month overhaul that included installation of the Naval Tactical Data System. She would serve as an operational test and evaluation (OPEVAL) ship.[6] At the end of testing the following spring, it was found that the equipment, which had deliberate redundancy and graceful degradation capability built in, had proved to be quite reliable, but that software programming problems were threatening the success of the effort. The system was given provisional service approval to allow the Navy to continue cleaning up the programs while equipment was built and data system technicians trained. The system would be installed on guided-missile frigates and attack carriers with the *America*, *Kitty Hawk*, and *Constellation* having high priority because they carried Terrier missile batteries.[7]

The Soviet Submarine Threat

Like the Americans, the Soviets after World War II had access to German submarine technology and eventually developed nuclear power for their growing undersea fleet along with anti-ship cruise missiles. Because the names of individual Soviet submarines were seldom known, they are usually identified by their NATO code names for each class, such as *Alfa*, *Charlie*, and *Kilo*. The first Soviet nuclear-powered submarines were the *November*-class fast attack submarines, which entered service in 1958. They were armed with torpedoes and, although they had a high underwater speed, were considerably noisier than diesel submarines and the early American nuclear-powered submarines. The *Victor* class, which could also launch cruise missiles, followed in 1979.[8] America responded to the growing undersea threat by developing specialized antisubmarine aircraft assigned to antisubmarine carriers (CVSs), usually converted *Essex*-class attack carriers, as well as land-based patrol planes, such as the Lockheed P2V Neptune and its successor, the P-3 Orion, supported by an extensive Sound Surveillance System (SOSUS) network.[9] By the early 1960s the *Essex*-class CVS ships had shifted their focus from mid-ocean protection of sea lanes to operating alongside the attack carriers for the protection of carrier task forces. (The Soviets had also developed their own ocean surveillance system to cue their attack submarines by this time.)[10]

The *Kitty Hawk* had been included in the FY56 budget, the *Constellation* in FY57, and the *Enterprise* in FY58. Because of the cost involved with the *Enterprise* and the desire to cut defense spending on the part of the Eisenhower administration, there were no carriers included in the FY59 and FY60 budgets. (The funds were, however shifted to the Polaris missile program.)[11] With the original *Forrestal*-class ships, there were features carried over from the original flush deck design that would prove to be drawbacks in operation and the follow-on carriers of the *Kitty Hawk* class corrected many of these shortcomings. Among the improvements over the original *Forrestal* design was the inclusion of more powerful C-13 steam aircraft catapults.[12]

Although they were similar in design, the biggest differences from the earlier ships, apart from their greater length, was a different placement of the elevators and the island. The forward port elevator on the *Forrestal* was moved aft of the portside catapults on the angle, which made it more useful for aircraft movement (the forward end elevator was rarely used anyway since it was in the landing path and the launch path of the waist catapults) and the island on the starboard side was moved aft so that two elevators were forward of the island and one aft, which increased the usefulness of the triangular parking area between the bow and the angle. The elevators were also enlarged by adding pie-shaped wedges to their outboard edges to accommodate anticipated growth in aircraft size. (This was to accommodate the A3J Vigilante, the last of the strategic attack bombers.)[13] With the ever increasing number of electronic systems on board aircraft carriers, the design for the island on the *Kitty Hawk* had become so crowded that a lattice mast behind the island was added to accommodate search radars and other electronics. This feature would be included on subsequent classes of aircraft carriers. Propulsion was provided by a 280,000 shp steam-turbine plant with four shafts, four Westinghouse steam turbines, and eight 1,200 psi Foster Wheeler boilers capable of driving these ships at 33 knots.

As designed, the first three ships were armed with Terrier surface-to-air missile systems. The Terrier was one of the Navy's earliest surface-to-air missiles (SAM). The two-stage, medium-range missile started out as a beam-riding system with a 10 nautical mile range at a speed of Mach 1.8, and ended up as a semi-active radar homing system with a range of 40 nautical miles at speeds as high as Mach 3. (It was replaced in service on surface combatants by the RIM-67 Standard Missile.)[14] Since the supporting missile launchers and the associated SPG-55 radars consumed a large amount of space, they were later removed when the Navy decided that the missiles duplicated the capabilities of the dedicated air defense escorts. Later in their operational careers the Terriers would be replaced by short-range systems to provide at least some self-defense capability—these would include the Sea Sparrow, the Close-In Weapons System (CIWS), and Rolling Airframe Missile (RAM). (These systems will be covered in detail in chapter 7.)

The *Kitty Hawk* was laid down at the New York Shipbuilding Corporation in Camden, New Jersey, on 27 December 1956 and launched on 21 May 1960 by flooding the drydock were she was built. Because of her bulk and the possibility of impact on the Philadelphia shore across the Delaware River, the conventional means of launching by sliding down the ways was ruled out. She was commissioned on 21 April 1961 at the Philadelphia Naval Shipyard and, following a shakedown in the Atlantic, *Kitty Hawk* departed Norfolk in August 1961 to transfer to the Pacific via Cape Horn with stops in Brazil, Chile, and Peru before arriving at San Diego in November. The "Hawk" would spend the rest of her career with the Pacific Fleet and would replace the *Independence* as the overseas home ported carrier in Yokosuka, Japan, in 1998. When she was decommissioned in May 2009, she was the last of the conventionally powered carriers in Navy service.

The *Constellation* was laid down at the New York Naval Shipyard in Brooklyn on 14 September 1957 and launched on 8 October 1960. (She would be the last carrier built by a shipyard other than Newport News.) While fitting out prior to her commissioning, a fire broke out on 19 December 1960 when a forklift on the hangar deck accidentally knocked over a steel plate, which then broke the plug off a 500 gallon tank of diesel fuel. At some point the diesel fuel, which had spilled from the tank and leaked down to the lower levels of the ship, was ignited (perhaps by a welder's blowtorch). The flames spread quickly and then moved to wooden scaffolding, filling the passageways of the ship with smoke. It took firefighters 17 hours to extinguish the fire, and although they saved hundreds of lives without losing any of their own, fifty shipyard workers had died. The extensive damage cost $75 million to repair and, after a delay of seven months, the *Constellation* was finally commissioned on 27 October 1961. After her fitting out and acceptance trials, *Constellation* left her home port of Norfolk in February 1962 for air operations off the Virginia Capes. After a month of local operations, "Connie" began a two-month shakedown cruise in the Caribbean. She departed from Mayport, Florida, in July for transfer to the Pacific by way of a two-month trip around Cape Horn. Operating out of her new home port of San Diego, she began working up prior to her first deployment to the Western Pacific from February to September 1963.

Following the *Kitty Hawk* and *Constellation*, the *Enterprise*, the world's first nuclear-powered carrier, was commissioned in November 1961. Originally intended to be the first of a class of six nuclear-powered carriers, her cost of construction had ballooned to over $450 million, which resulted in her becoming the only ship of her class. In order to cut costs, she was also completed without her planned Terrier missile armament.[15] At the time, the operational benefits of nuclear power for aircraft carriers were not clearly understood and it was difficult for advocates of nuclear power to make their case without actual experience to back them up. Since the Navy had regarded the *Enterprise* as an experimental ship it was decided that the *America*, which would also be built at Newport News Shipbuilding, would be conventionally powered. *America* was laid down on 9 January 1961, launched on 1 February 1964 under the sponsorship Mrs. David L. McDonald (wife of Admiral David L. McDonald, then the Chief of Naval Operations), and commissioned at the Norfolk Naval Shipyard on 23 January 1965. (Apparently the decision to build the *America* with conventional propulsion was not universally accepted. On 18 November 1964 when the *America* left Newport News with a civilian crew on her sea trials, her commanding officer, Captain Lawrence Heyworth Jr., sent a message to Vice Admiral Paul R. Ramsey, Commander, Air Forces Atlantic [ComAirLant]: "Under way on fossil power.")[16] After fitting out at Norfolk, *America* briefly operated off the Virginia Capes before conducting her shakedown in the Caribbean. She entered the Norfolk yard for post-shakedown upkeep and after operating locally for some months, sailed for her first Mediterranean deployment late in 1965.[17] Although built to the basic *Kitty Hawk* design, the *America* had several differences from the lead units of her class. Instead the normal arrangement of two bow anchors, port and starboard, *America* had no starboard anchor and another anchor astern. This change was made to accommodate the AN/SQS-23 sonar. *America* was the only American carrier built after World War II with sonar, although it was removed in the early 1980s. (Anti-submarine warfare [ASW] carriers of the *Essex* class had sonar installed as part of their Fleet Rehabilitation and Modernization [FRAM] overhauls.) The reason for the sonar was to provide a self-defense capability against the newer Soviet nuclear-powered attack submarines, which could keep up with the fast carrier formations.[18] She also had a narrow smokestack compared to prior units.

The ongoing debate over the costs versus benefits of nuclear ships would affect later decisions over what would eventually become the *John F. Kennedy*.[19] Although the *John F. Kennedy* was similar to the earlier units in the arrangement of the flight deck and in its propulsion system, there were enough differences that she is often placed in her own class. The Navy had prepared various design alternatives for both nuclear-powered and conventionally powered carriers, but a nuclear-powered carrier with comparable capabilities still would cost one third to one half again as much its conventionally powered counterpart. Moreover, at that point no nuclear-powered ship had yet been evaluated under actual service. When the Kennedy administration came into office in January 1961, there was yet another review of the carrier program and carriers were being planned for every other fiscal year after FY63. In 1963 Defense Secretary Robert McNamara ordered the cancellation of the FY65 ship. By then, operational experience with the *Enterprise* and the nuclear-powered cruiser *Long Beach* had convinced the Navy of the value of nuclear power for surface ships, but McNamara rejected the Navy's recommendations and CVA-67 was ordered as a "fossil-fuel" carrier in October 1963, about a year behind the FY63 schedule. His decision led to protest from the Navy and within Congress and Navy Secretary Fred Korth resigned (in an action similar to that of Navy Secretary John Sullivan's reaction when the *United States* was canceled in 1949).[20] The keel was laid on 22 October 1964 by Newport News Shipbuilding and the ship was officially christened on 27 May 1967 by Jacqueline Kennedy and her nine-year-old daughter, Caroline, two days short of what would have been President Kennedy's fiftieth birthday.[21] The design was to have included a bow-mounted sonar and a Tartar missile battery, but she was completed without these, although the bow-mounted anchor similar to that of the *America* was retained.[22] The island differed from the *Kitty Hawk* class, with angled funnels to direct smoke and gases away from the flight deck, which had a slightly different outline to the port side at the front end of the angle (similar to the later *Nimitz*-class carriers), and she was also 17 feet shorter. Perhaps most significant, however, was the different internal hull design. In previous carriers the fuel oil stored in voids along the side of the hull structure formed part of the protection for vital machinery spaces. With a nuclear-powered carrier this space could be used for additional aviation fuel, or, in order to increase internal volume for other uses and thus keep the overall hull size within limits, a more compact arrangement of voids could be provided. Ironically, the new system developed for the nuclear-powered CVAN-67 alternative was incorporated into the non-nuclear ship that was in fact built.[23] She was commissioned on 7 September 1968 and conducted her initial shakedown runs along the eastern seaboard. After completing her operational readiness inspections,

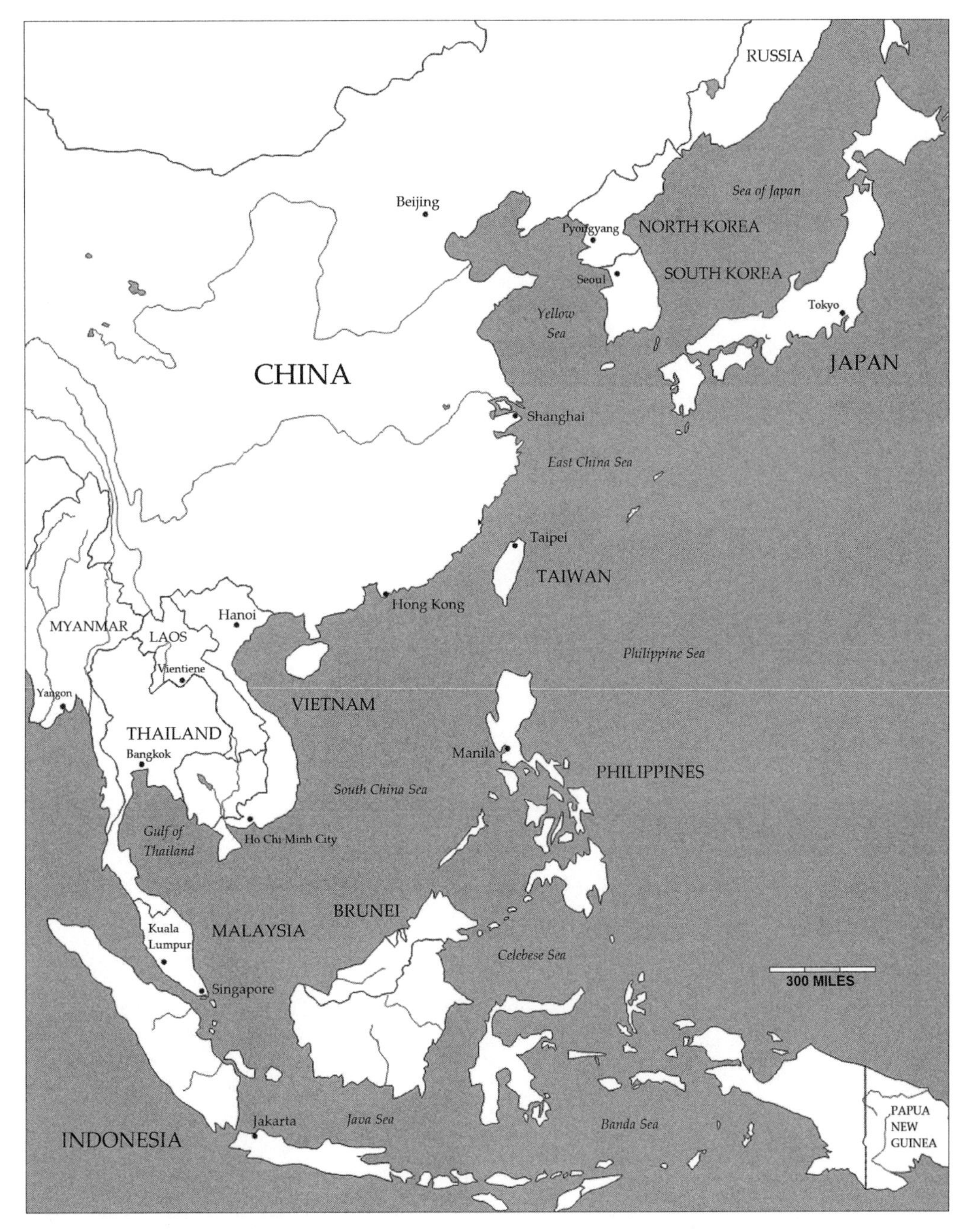

Map 1. The Western Pacific Map adapted by the author from (East Asia) http://d-maps.com/carte.php?num_car=28778&lang=en.

"Big John" departed in April 1969 to relieve the *Forrestal* on the first of many deployments to the Mediterranean.

Once experience with the effectiveness and availability of carrier-based tactical air power in Vietnam became evident, Secretary McNamara, in testimony to Congress in 1966, advocated a force of 15 carriers and 12 air wings. The FY67 budget would include CVAN-68, which would eventually be named *Nimitz*, the first ship in a class of ten nuclear-powered carriers. The *Kennedy* would be the last conventionally powered carrier to be built for the U.S. Navy.[24]

Cold Wars and Hot Spots 5

In the late 1950s and early 1960s, even as critics of the Navy's carrier construction plans discounted the need for carriers and pointed to their assumed vulnerability to the growing threat of supersonic aircraft and missiles, carrier forces were often called upon to support the American policy of containment of communist aggression during the growing Cold War with the Soviet Union and its allies and client states. These crises were brought into focus by a statement by Soviet Premier Nikita Khrushchev in 1961 when he stated, "There will be liberation wars as long as imperialism exists. Wars of this kind are revolutionary wars. Such wars are not only justified, they are inevitable. . . . The peoples win freedom and independence only through struggle. . . . We recognize such wars. We have helped and shall continue to help people fighting for their freedom."[1]

Although the official American and North Atlantic Treaty Organization (NATO) policy was that any communist encroachment on the free world would be met with swift retaliation by U.S. nuclear bombers, conventional military forces and economic aid had stopped communism in Western Europe, Greece, Turkey, Lebanon, Malaya, the Philippines, and Korea. Since aircraft carriers could quickly provide an American presence in various hot spots around the globe, merely moving a carrier task force into an area demonstrated American capability and resolve, yet at the same time did not interfere with any nation's sovereignty or commit American forces to any particular course of action. As the 1960s opened the Navy had a force of 14 attack carriers: four *Forrestal*-, three *Midway*-, and seven *Essex*-class ships. The attack carriers were supplemented by nine *Essex*-class antisubmarine warfare (ASW) carriers with a tenth operating as a training carrier.[2]

The Middle East

American sea power has frequently been used as a stabilizing influence in the volatile Middle East, where America has long sought to balance its support of Israel against relations with other Middle Eastern states, often in the face of Soviet efforts to dominate the region during the height of the Cold War through Arab client states and to deny Western access to Arab oil. It was typical for the Navy to deploy one large carrier, either a *Forrestal* or *Midway* class, and one or two of the smaller *Essex*-class carriers with the Sixth Fleet in the Mediterranean.

The Suez Crisis

In July 1956 Egyptian president Gamel Abdel Nasser decided to nationalize the Suez Canal, after Britain and the United States withdrew their offers to fund the building of the Aswan dam, partly in response to Nasser's ties to the Soviet Union and his recognition of China at the height of tensions between the Republic of China and Taiwan. As the only direct route from the Mediterranean to the Indian Ocean, the Suez Canal is vital to the flow of trade between Asia, the Middle East, Europe, and the U.S. Angered by Nasser's action, Britain and France joined Israel in an attack on Egypt on 29 October, quickly defeating the Egyptian army and occupying the Suez region. The Egyptians responded by blocking the canal. The United Nations intervened and President Eisenhower came out strongly against the military intervention by two of America's closest allies. Britain and France backed down, giving control back to Egypt the following March. In the meantime, President Eisenhower was prepared to use force if necessary. The *Forrestal*, having spent her first year operating off the Virginia Capes and in the Caribbean, left Mayport on 7 November 1956 to operate in the eastern Atlantic, ready to enter the Mediterranean should it be necessary.[3]

Lebanon

In April 1957 the Sixth Fleet operated in the Eastern Mediterranean as a show of force in support of Jordan's King Hussein, who had just put down a revolt by pro-Egyptian, communist-inspired elements in the Jordanian army. In February 1958 Nasser's Egypt had merged with Syria to form the United Arab Republic (UAR), which became a threat to Jordan and Iraq. (The UAR dissolved in 1961 when Syria left, but Egypt continued to be known as the UAR officially until 1971.) Unrest in the form of armed rebellion spread to Lebanon in early May 1958 and when the pro-Western government of Iraq was overthrown by a coup in July 1958, Lebanese president Camille Chamoun asked for American assistance in preventing a similar occurrence in his own country. Within hours naval forces of the Sixth Fleet went into action as part of Operation Blue Bat. At the time the *Saratoga* was at Cannes, France, some 1,700 miles away, while the *Essex* was at Piraeus Bay off Athens, Greece, and the ASW carrier *Wasp* was at Naples, Italy. (The *Forrestal* was also again ordered to the eastern Atlantic to be ready to back up naval operations in the Mediterranean if needed.) The Marines had been given the order to land on 15 July, to coincide with President Eisenhower's announcement of American intervention. The carriers had recalled their crews from shore leave the night before and sped to the east. Other American forces in Europe could not respond in time, but one battalion of Marines quickly moved ashore and secured Beirut. Fortunately, the landing, covered by aircraft from the *Saratoga* and *Essex*, was unopposed as crowds of Lebanese greeted the Marines at the beach. The Beirut airport was secure within an hour and more battalions of Marines were soon landed.

Although the landings were peaceful, the American capability for rapid intervention did not go unnoticed by either the Soviet Union or the more radical Arab states in the region, such as Nasser's Egypt. (The British had also responded to an appeal from Jordan's King Hussein with 2,000 airborne troops and French naval forces had appeared briefly off Beirut.) By late August the Middle East situation was relatively stable and the last of the 6,000 Marines and 8,000 soldiers who had gone into Lebanon were withdrawn in late October.[4] In the aftermath of the 1967 Arab-Israeli War, problems would again arise in Lebanon as Palestinian refugees and thousands of Palestinian militiamen from Yasser Arafat's Palestine Liberation Organization (PLO) regrouped in southern Lebanon, with the intention of attacking Israel. Beginning in 1968, these attacks eventually provoked an Israeli response. Although these reprisals were intended to induce the Lebanese government to crack down on the PLO, they served only to polarize the various factions in Lebanon, eventually leading to civil war in 1975. Occasionally, tensions would require the carriers to stand by for contingencies. In May 1973, the *John F. Kennedy* and *Forrestal* were alerted for possible contingencies when fighting broke out in Lebanon between Lebanese army units and Palestinian guerillas in Lebanon. Fortunately, a cease-fire was arranged.[5]

Africa

The Middle East was not the only area of concern as the 1960s opened. In mid-1960 civil war erupted in the newly independent Congo and in July the ASW carrier *Wasp* was rushed to Ghana to unload aviation gasoline for transport aircraft flying troops and supplies for the U.N. peace-keeping forces operating in the Congo.[6] In the Horn of Africa, Somalia was created in 1960 from a former British protectorate and an Italian colony. In 1969 Mohamed Siad Barre seized power and established the Somali Democratic Republic as a socialist state the following year. Following a Marxist model, almost all industries, banks, and businesses were nationalized, and cooperative farms were promoted. Over the next eight years, the military relationship with the Soviet Union grew. Somalia signed an agreement to improve and modernize the port of Berbera in return for Soviet access to the facility. Eventually Berbera included a missile storage facility for the Soviet navy, an airfield with runways capable of handling large bombers, and extensive radar and communications facilities. Access to Berbera gave the Soviet Union a presence in the strategically important Indian Ocean–Persian Gulf region to counter U.S. presence in the area.

Berbera acquired even greater importance when Egypt expelled all Soviet advisers in July 1972. In 1977 Siad Barre ended the friendship treaty with the Soviet Union and expelled Soviet advisers.[7] Siad Barre then switched allegiance to the West and the U.S. provided significant aid to Somalia until 1989. Since Siad Barre was overthrown in 1991, Somalia has been without an effective central government and would again become a hot spot requiring American involvement in the coming decades.

The Six-Day War and the *Liberty* Incident

On the morning of 5 June 1967, while the *America* was refueling from the oiler *Truckee*, word came that the Israelis and the Arabs were at war. What later became known as the "Six-Day War" began when Israel launched surprise raids on Egyptian airfields after tensions with Egypt, Jordan, and Syria had escalated over the previous months. The high-tempo conflict saw Israeli raids on the West Bank, aerial encounters over Syrian territory, clashes with Syrian forces in the Golan Heights, a naval blockade by Egypt, and the evacuation of the U.N. peace-keeping force in the Sinai. On 8 June, in mid-afternoon, the American intelligence-gathering ship *Liberty* was attacked by Israeli torpedo boats and aircraft while in international waters about 15 miles north of the Sinai port of El Arish.[8] In a matter of minutes, F-4 Phantoms were launched to defend against any possible attack against the carrier task force. Also, four A-4 Skyhawks were loaded and launched to come to *Liberty*'s aid. As the Skyhawks and their fighter escort flew toward *Liberty*'s position, word was received that the attackers were Israeli and that the attack had been made in error. The outbound planes were recalled with ordnance still on their racks. On the *Liberty* 34 men had been killed and 75 wounded, 15 seriously. On the morning of 9 June, the *Liberty*, listing and perforated by rockets and cannon shells, came alongside the *America*. As she drew near the flight deck crew, moved by the battered ship's appearance, gave her crew a resounding cheer.[9] (The controversy over this Israeli attack on the *Liberty*, which many believed was *not* a mistake, continues to this day.) By 10 June Israel had won a decisive victory—Israeli forces had taken control of the Gaza Strip and the Sinai Peninsula from Egypt, the West Bank and East Jerusalem from Jordan, and the Golan Heights from Syria. Israel had demonstrated that it was willing, and able, to launch strategic strikes that could change the regional balance of power, but its enemies in the Arab world were determined to avenge their humiliation.

Yom Kippur War

In the years following the Six-Day War, Israel fortified both the Sinai and the Golan Heights. But the stunning Israeli victory had taught Egypt and Syria some hard tactical lessons and they would launch an attack in 1973 in an unsuccessful attempt to reclaim their lost territory. From the end of 1972, Egypt began building up its forces with Soviet-supplied MiG jet fighters, antiaircraft missiles, tanks, and antitank weapons. Political generals were replaced with competent officers and military tactics, based on Soviet battlefield doctrines, were improved. Egypt also engaged in strategic deception, holding large-scale mobilization and training exercises intended to convince the Israelis that the indications of a military buildup they had observed were not preparation for an attack. In the afternoon of 6 October 1973, during the Jewish holiday of Yom Kippur, Egypt launched a large air strike by over 200 aircraft against

Israeli air bases, surface-to-air missile batteries, command centers, artillery positions, and radar installations. In preparation for the assault across the Suez Canal, Egypt had deployed five divisions with 100,000 soldiers, 1,350 tanks, 2,000 guns, and heavy mortars. Anticipating an Israeli armored counter attack, the assault forces were well equipped with antitank weapons, and the Egyptian air defenses blunted retaliatory Israeli air strikes. In the north, Syrian forces simultaneously attacked the Golan Heights. After initial setbacks, the Israelis regained lost territory and even managed their own crossing to the west bank of the Suez Canal. In response to the fighting, on 8 October the Navy ordered two carrier battle groups centered on the *Independence* and *Franklin D. Roosevelt*, along with the helicopter carrier *Guadalcanal*, to stand by for evacuation operation contingencies until 13 October, with the *Independence* operating off Crete.

Quemoy and Matsu

While American attention was drawn to the Middle East, another situation was developing in Asia. On 23 August 1958, communist Chinese artillery on the mainland opened fire on the offshore islands of Little and Big Quemoy, Nationalist Chinese strongholds garrisoned with 60,000 troops. Another 25,000 were on Matsu Island to the north, which was also shelled. The intense bombardment threatened to cut the islands off from supplies and ammunition from Formosa (as Taiwan was often referred to in those days). At the time, the *Hancock* was operating east of Formosa, the *Shangri-La* was at Yokosuka, Japan, and the *Lexington* and the ASW carrier *Princeton* were at sea east of Japan. The U.S. government had declared that it would defend Formosa against a communist Chinese attack, but the degree of American commitment to defend the offshore islands was uncertain. In view of the threat, however, the Seventh Fleet was placed on alert and prepared contingency plans for major operations against communist China. The *Midway* had left the West Coast the week before for her deployment to the Western Pacific (WestPac). She was at Pearl Harbor on the 23rd and was ordered to join Task Force 77 immediately. Task Force 77 was also joined by the *Essex*, which had transited the Suez Canal from the Mediterranean after supporting the landings in Lebanon. The rapid buildup of the Seventh Fleet and of other U.S. forces on Formosa convinced the communist Chinese of U.S. determination to help the Nationalist Chinese hold the islands. The intensive fire against the islands was lifted, but resumed on alternating days with much less intensity.[10]

In 1961, as the need for U.S. naval strength in the Mediterranean and Western Pacific increased, events in the Caribbean caused the Navy to deploy ships to that area. Besides the ill-fated Bay of Pigs invasion in April, the *Intrepid*, *Shangri-La*, and *Randolph* were ordered to stand by off southern Hispaniola when a general uprising appeared imminent following the assassination of President Molina Trujillo of the Dominican Republic in May.[11] In November 1961 carrier forces patrolled off Guatemala and Nicaragua when their governments requested U.S. assistance in preventing a communist-led invasion of those nations.[12]

The Bay of Pigs

When John F. Kennedy took office in 1961, he inherited a plan from the Eisenhower administration for the invasion of communist Cuba, taken over by Fidel Castro in 1959. The plan called for Cuban exiles, transported and supplied by the U.S. government, to liberate Cuba by landing at the Bay of Pigs, located on the southeastern shore of Cuba. Although the administration planned and supported the operation, great pains were taken to conceal official involvement. The *Independence* and *Essex* were ordered to the area in April, but the rules of engagement prevented any direct interference by American naval forces unless actually fired upon. At a critical point in the invasion, when U.S. air support was vital to the survival of the invaders, the *Independence* and *Essex* could only stand by helplessly as the Cuban exiles were defeated, rounded up, and either killed or captured.[13]

The Cuban Missile Crisis

Throughout the summer of 1962 there were indications of unusual activity in Cuba. In October high-altitude reconnaissance flights by U-2 spy planes showed that the Soviets were constructing missile bases in Cuba. Along with the Il-28 twin-engine jet bombers also arriving in Cuba, these missiles could reach targets in the United States within minutes. Faced with a choice of ignoring the threat or starting a war between the two super powers, President Kennedy opted for a naval blockade, euphemistically called a "quarantine." Kennedy announced the quarantine on national television on 21 October. To give the Russians time to think things over, the blockade would not actually go into effect until the morning of 24 October. Task Force 135, with some 483 ships deployed in an arc about 500 miles to the northeast of Cuba, stood ready. Nikita Khrushchev, the Soviet premier, called the American allegations of missile bases in Cuba lies and threatened that if the American navy carried out any "act of piracy" the Soviet Union would react accordingly. Tension mounted as the Russian cargo ships en route to Cuba approached the American naval forces. Fortunately, Khrushchev backed down and the crisis abated. The Russians dismantled their missile bases and removed them from Cuba.[14]

Cold War Encounters at Sea

Soviet reconnaissance aircraft had occasionally overflown U.S. carriers, but beginning in late January 1963 the Soviets began a series of carrier overflights, in what many regarded as a move on the part of Admiral of the Fleet and Chief of the Soviet Navy, Admiral Sergey Gorshkov, to counteract the humiliation the Soviet Union experienced during the Cuban missile crisis of the previous year. The flights were probably meant to demonstrate not only the vulnerability of American carriers, but also the growing Soviet capability for "blue water" operations. Between 27 January and 27 February, the Soviets overflew the *Enterprise* and *Forrestal* in the Atlantic and the *Constellation*, *Kitty Hawk*, and the helicopter carrier *Princeton* in the Pacific. (The

Constellation was nearly 600 miles south of Midway Island when she was overflown; the *Kitty Hawk* was in the North Pacific.)

The experience of the *Forrestal* was typical of these early incidents. While returning from a deployment in the Mediterranean, she was just south of the Azores. At the time she was not conducting flight operations (crewmen were even on the flight deck flying kites), when a large Soviet bomber approached from the north and flew directly over the *Forrestal* at about 200 feet. The Bear, as it was later identified, then circled around and made another dangerous low-level pass before heading back to its base in Murmansk, thousands of miles to the north.[15] (Bear is the NATO code name for the Tupolev Tu-95, a large swept-wing, four-engine turboprop bomber that was contemporary with the American B-52. Like the B-52, it was conceived as a strategic bomber but continued to serve long after the aircraft intended to replace it had come and gone, mainly because of its adaptability to other roles.)[16]

In response, Defense Secretary McNamara, on 28 February 1963, revealed the overflights to the press, noting that any nation had the right to photograph ships in international waters and that he saw no cause for alarm. Navy officials also quickly pointed out that the carriers were on normal sailing routes with published schedules and were not operating under radio silence at the time. In June six Badger jet bombers overflew the *Ranger* 330 miles east of Japan.[17] (The Tupolev Tu-16 Badger was a swept-wing twin-engine jet bomber, comparable in many respects to the American B-47, that entered Soviet service in the mid-1950s. It was adapted to many roles and was built under license in China as the Xian H-6. The Soviets also exported Tu-16s to Egypt, Indonesia, and Iraq.) In all, there were 14 overflights in 1963, dropping to four in 1964, after which the Soviets returned to their usual practice of intelligence gathering by assigning "fishing trawlers" (AGIs) to shadow carrier task groups.[18] However, overflights would continue sporadically over the years. Badgers and Bears flew from the Soviet Union and, after March 1968, Badgers in Egyptian markings were flown by Soviet crews from Cairo West airfield in Egypt. That these confrontations were potential disasters waiting to happen was obvious to participants on both sides. An example was the 25 May 1968 overflight by a Soviet Badger over a group of U.S. Navy vessels, including the *Essex*, off the coast of northern Norway. After a low pass over the *Essex*, the Soviet bomber banked away sharply. One wing tip hit the sea and the plane cartwheeled and exploded. There were no survivors. In another case, a Soviet Badger overflew the *John F. Kennedy* while she was operating in the Norwegian Sea on 4 October 1973. An F-4 attempting to escort the bomber away from the area collided with it. The Badger returned to its base and the F-4 landed in Norway.[19]

The Navy at first responded by changing its operational procedures. Transits would be planned to avoid the normal "great circle" routes, eliminate electronic emissions (EMCON Alpha—no radio communications, radar, IFF, etc.) and "darken ship" (showing only the navigation lights required for safety) when crossing merchant sea lanes at night. While this resulted in no encounters with Soviet aircraft, or ships or submarines for that matter, on further consideration, the Navy could not be sure that they had indeed avoided detection or that the Soviets had only decided not to respond. Changing its approach, the Navy tried something called the "*Kitty Hawk*

Express." When the *Kitty Hawk* deployed to the Far East in 1965, she followed the usual procedures to avoid detection until she was within range of Soviet bomber bases on the Kamchatka Peninsula. When she resumed normal high frequency communications, signals intelligence (SIGINT) stations in Japan noted an immediate reaction at bomber bases on Kamchatka and Soviet bombers were launched within minutes. They were soon flying over the *Kitty Hawk*, but not before they had been intercepted by F-4 fighters at about 150 nautical miles from the carrier, beyond the bombers effective weapons range, and "escorted" all the way in.[20]

As the Cold War heated up in the mid- to late 1960s, potentially dangerous, life-threatening incidents involving ships and submarines, as well as aircraft, became all too common. In 1968 the U.S. proposed talks with the Soviet Union on reducing the chances for incidents between the two countries to occur, and to prevent them from escalating if they did occur. The Soviets agreed and after talks in Moscow and Washington, the Incidents at Sea (INCSEA) Agreement was formally signed in Moscow in March 1972 by Navy Secretary John Warner and Soviet Admiral Sergey Gorshkov.[21] Although the agreement set rules for both parties, the games continued through the years. The idea was to intercept Soviet aircraft far enough away so as to demonstrate that they could be destroyed before being able to attack the carriers, but not so far out that it might give away the ability to detect incoming aircraft. The boundaries were set between 150 to 200 miles out. (In a study done by the Center for Naval Analyses of approximately 150 cases of U.S. aircraft carriers vs. Soviet naval aviation from 1963 to 1975, it was shown that carriers were not as vulnerable as the Soviets, or critics in the U.S. Air Force for that matter, believed. The results did show, however, that naval forces had to be vigilant and situation sensitive, and make use of all available assets—surface ships, aircraft, and attack submarine escorts.)[22]

An overflight by Soviet aircraft was often the highlight of a carrier's transit from ports on the east or west coasts to operating areas in the Atlantic, Mediterranean or the Pacific. Photo planes would often accompany the fighters to take pictures of them flying in formation with a Bear or Badger. At times the Soviet air crews would be taking pictures of their own and sometimes holding up signs or even Playboy centerfolds.

Encounters with Soviet submarines were also a cause for concern. For example, in March 1984, as the *Kitty Hawk* battle group headed south toward the Tsushima Strait into the Yellow Sea during Exercise Team Spirit, a Soviet submarine, identified as a probable *Victor*-class attack boat, collided with the carrier as it was surfacing. The collision occurred about 150 miles east of Pohang, South Korea. On the bridge, the Hawk's captain felt a noticeable shudder and the outline of the sub's sail was seen moving away to starboard. Helicopters from the Hawk inspected the sub with night vision goggles and reported no serious visible damage. Although the sub did not display any navigation lights, it apparently remained seaworthy. As the carrier and her screen stood by to render assistance, an attempt was made to contact the cruiser *Petropavlovsk*, flagship of the Soviet task force shadowing the American battle group, by flashing light, but got no response.[23]

But the deadly seriousness behind the game remained as each side still vied to hone their skills and gather intelligence, resulting in occasional encounters, even

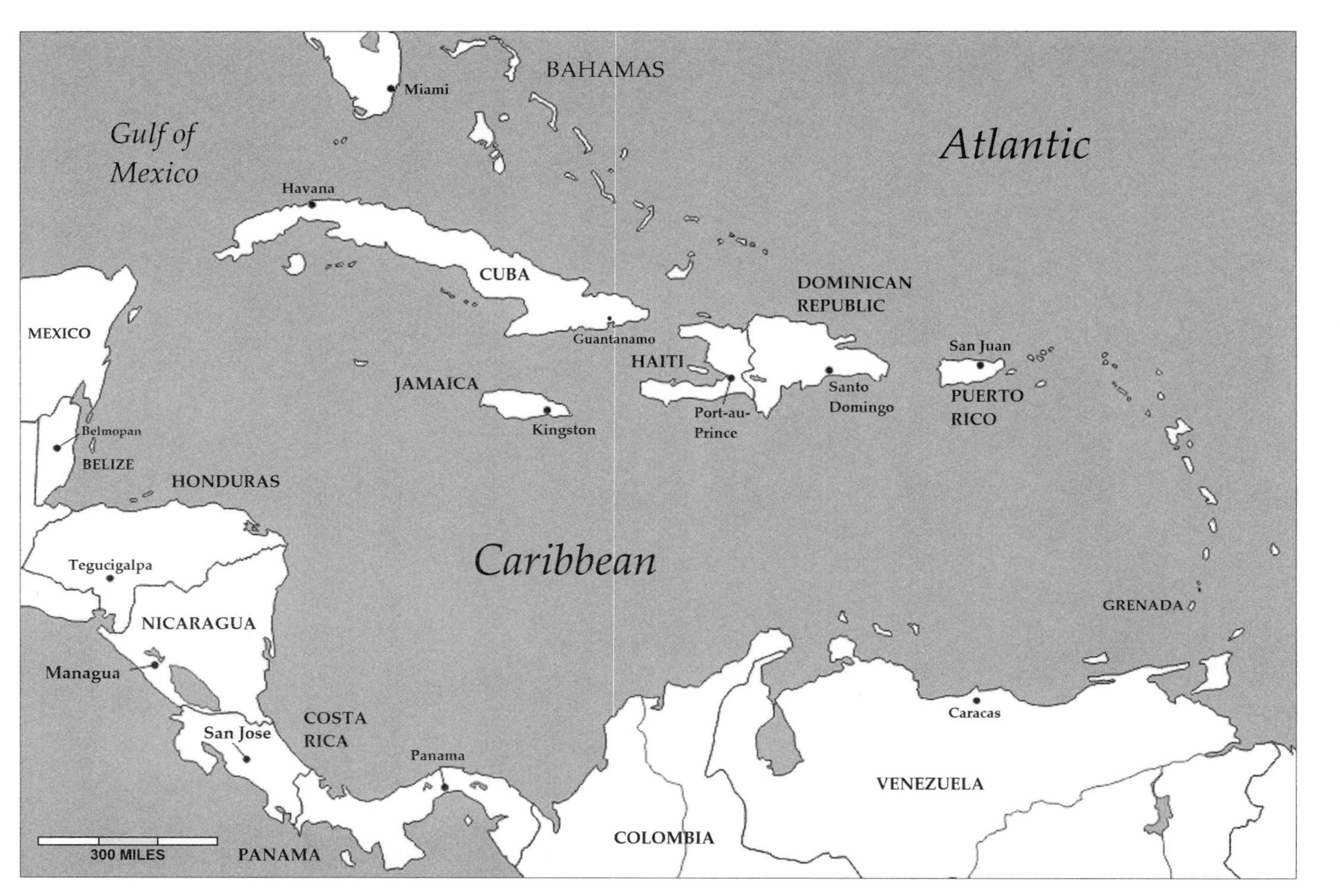

Map 2. The Caribbean

Map adapted by the author from (Central America) http://d-maps.com/carte.php?num_car=22802&lang=en.

after the fall of the Soviet Union. On 17 October and again on 9 November 2001, while operating in the Sea of Japan, the *Kitty Hawk* was, according to the Russian air force commander, caught by surprise. Their air crews took pictures of the flight deck crew's reaction and the Russians later e-mailed them to the Hawk's skipper. (The *Kitty Hawk* was conducting underway replenishment [UNREP] when she was overflown twice on 9 November and could not launch aircraft to escort the Russian Su-24MR Fencer reconnaissance aircraft and accompanying Su-27 Flanker fighters.)[24]

Next-Generation Aircraft 6

In the early 1960s new carrier aircraft were designed to take advantage of the potential offered by the super carriers. These next generation carrier aircraft would greatly increase the capabilities of naval aviation and their performance would often equal or surpass that of their land-based Air Force contemporaries. With the new administration of John F. Kennedy, which began in January 1961, Robert McNamara, as Secretary of Defense in the Kennedy, and later, Johnson administrations, adopted a systems analysis approach to weapons acquisition. Under McNamara, commonality was viewed as one means for achieving greater cost effectiveness while eliminating duplication of capabilities between the services. Unlike previous defense secretaries, he tended to rely more on statistics than on the advice of military leaders, which often generated controversy between and within the services. A major area of McNamara's cost reduction efforts was the consolidation of programs from different services, most notably in aircraft acquisition, and one of the implications of this effort was the adoption in 1962 of a common designation system between the services. Under the old Navy system, both the mission and manufacturer of an aircraft were represented. The F8U Crusader, for example was the eighth fighter ("F") design by the Vought ("U") company for the Navy; successive versions were identified by numbers following the basic designation and letters were used to denote special roles (as shown by the photographic version of the Crusader, the F8U-1P). In the late 1950s and early 1960s, many of the Navy aircraft developed under the old system were later redesignated in 1962 in the new system, which was basically that used by the Air Force. Many of the basic functions were the same, "F" for fighter, "A" for Attack, etc., but the numbers were in sequence regardless of the manufacturer. Special roles were identified by prefix letters (such as "R" for reconnaissance) and successive versions were given letters (usually, but not always in alphabetical sequence.) The Crusader became the F-8 and the photo version would become the RF-8A. (A list is included in appendix C.) Beyond the administrative changes in aircraft designations, there was an increased emphasis on the same aircraft being used by more than one service if they performed similar roles. This worked to the Navy's advantage in some cases, but in others the Navy was forced to accept an aircraft that was not appropriate.[1]

Phantom II

The McDonnell F4H Phantom II was one of the most successful American fighters since World War II. Over 5,000 were built from 1959 until 1979 and many served with several foreign air forces.[2] Named the Phantom II to honor the first pure jet fighter developed for the Navy by McDonnell, the FH-1, most know it as *the* Phantom, since its ancestor was produced in limited numbers and served only briefly on active service.[3] The Phantom design started out as an improvement over the F3H Demon, and McDonnell developed single and twin-engine proposals.

Although designated as a fighter the emphasis was on attack capability. Meanwhile, the Navy, which was accepting its first Crusaders, decided that what it really needed was an all-weather interceptor armed with the new Sparrow air-to-air missiles. (At this time cannons for air-to-air combat were thought to be anachronisms. Later combat experience in Vietnam would prove the fallacy of such assumptions.) McDonnell reworked the design as a twin-engine, two-seat aircraft, and the XF4H-1 first flew in May 1958. Carrier trials began in the fall of 1959 and at the end of 1960 the Pacemakers of VF-121 became the first Phantom squadron.[4] Overall, the Phantom was a big brute of an aircraft. At first, many Navy and Marine aviators thought it was a "triumph of thrust over aerodynamics." (Its General Electric J-79 jet engines produced 10,000 pounds of thrust each, 17,000 pounds in afterburner.)[5] They called it the "rhino" and "double ugly" (also the "smoker" because of the trail of smoke left by its J-79 engines), but they soon learned that the Phantom had many virtues as the Navy began setting all kinds of aviation records with its new Mach 2.2 fighter. The VF-74 Be-Devilers became the first squadron to deploy with the Phantom. Having completed carrier qualifications in October 1961, VF-74 participated in the *Forrestal*'s August 1962 to March 1963 cruise to the Mediterranean.[6] Defense Secretary McNamara directed the Air Force to evaluate the Navy fighter in 1961 and the Phantom, even though it carried the weight of additional equipment for carrier operations, was more than a match for existing Air Force fighters. The Air Force would eventually acquire over twice as many Phantoms as the Navy and Marine Corps combined. In the 1962 designation scheme the F4H became the F-4, with the F-4B as the Navy version and the F-4C for the Air Force.[7]

Vigilante

The North American A3J Vigilante was a carrier-based supersonic nuclear strike bomber intended to replace the Douglas A3D Skywarrior. (It never did replace the A-3, which had been adapted for electronic warfare and tanking missions.) At the time it entered service in 1961, the Vigilante was one of the largest and most complex aircraft ever to operate from a carrier. It had a high-mounted swept wing with a boundary-layer control system (blown flaps) to improve low-speed lift, and there were no ailerons (spoilers and the all-moving tail surfaces provided roll control). It had two widely spaced GE J-79 turbojet engines (the same engines used by the F-4 Phantom), and a single large all-moving vertical stabilizer. The wings, vertical stabilizer, and the nose radome all folded for carrier stowage. The Vigilante had a crew of two seated in tandem, a pilot and a bombardier-navigator (BN) (reconnaissance/attack navigator [RAN] on later versions) in individual ejection seats. Although surprisingly fast and agile for an aircraft of its size, the high approach speeds and angle of attack of the Vigilante required a great deal of pilot skill in recovering on a carrier. Also, the advanced and complex electronics made it difficult to maintain. In the original attack version, a nuclear weapon and two disposable fuel tanks were carried in a "linear bomb bay" between the engines. The idea was for the weapon to be ejected rearward along with two disposable fuel tanks, which in practice was not a reliable method. No live weapons were ever carried in the linear bomb bay.

The Vigilante's role as a nuclear strike aircraft was short lived. The Navy's policy had shifted in favor of the Polaris submarine-launched ballistic missile rather than manned bombers. The "Vigi" found new life as a reconnaissance aircraft, serving throughout the Vietnam War and into the end of 1979. (The "straight" versions were canceled in 1963.)[8] The RA-5C had slightly greater wing area and added a long canoe-shaped fairing under the fuselage for a multi-sensor reconnaissance pack with side-looking airborne radar (SLAR), an infrared scanner, and camera packs, as well as improved electronic countermeasures (ECM). It also had television cameras and high-intensity strobe flashers for night photography. Eight of the ten Reconnaissance Heavy Attack Squadrons (RVAHs) served extensively in Vietnam starting in 1964 with the RVAH-5 Savage Sons on board the *Ranger*.

The RA-5C had the dubious honor of having the highest loss rate of any Navy aircraft in combat because the Vietnamese knew there would be post-strike bomb damage assessment (BDA) flights, and the ten minutes or so needed for the dust and smoke to clear was ample time for their gunners to reload and get ready.[9]

RVAHs originally started with six aircraft, but as attrition took its toll, this was eventually reduced to three. The aircraft itself was only the airborne part of an overall system known as the Integrated Operational Intelligence Center (IOIC).[10] All the reconnaissance information collected by the RA-5C was recorded on magnetic tape, which along with the photography, was immediately processed after recovery and the "read out" provided to other ships equipped with the same data link capabilities as the IOIC carrier.[11]

Intruder

In the 1950s the Navy began thinking about a medium attack aircraft to replace the AD Skyraider and complement the heavy attack A3D Skywarrior and the light attack A4D Skyhawk. Drawing on the lessons learned from night attacks in the Korean War, the Navy wanted a true all-weather attack capability at low altitudes and the A-6 Intruder became the first aircraft in history designed specifically to strike targets obscured by bad weather or darkness. Grumman won the contract to develop the A2F-1, as the Intruder was then known, and the first flight occurred in 1960. The Intruder had a tadpole-shaped fuselage, with a broad nose to accommodate the side-by-side cockpit and the radar system, that tapered back to a slender tail. It was powered by twin Pratt & Whitney J52 jet engines mounted against the fuselage, under the wing roots. The shoulder-mounted wings were swept back at 52 degrees and were optimized for subsonic speeds, which allowed the Intruder to carry a significant bomb load on its five pylons (four under the wings and a fifth station under the fuselage which normally carried a drop tank). The Intruder was initially designed with exhaust tailpipes that tilted downward to provide some short takeoff capability, but this feature was not included on production aircraft. (The original intent was to allow Marine Intruders to operate from shorter runways ashore.) Early versions of the Intruder had perforated speed brakes behind the jet exhaust, but these were soon wired shut and deleted on later production aircraft. The Intruder was also equipped with wingtip decelerons, a type of airbrake with two panels that open in opposite directions.

For its day, the Intruder had surprisingly sophisticated avionics with a high degree of integration. The first production model, the A-6A, entered squadron service in 1963. It was equipped with the Digital Integrated Attack/Navigation Equipment (DIANE) system that combined search and track radars; navigation, communications, and identification equipment; a cockpit display system; and a high-speed digital computer. The DIANE system enabled the pilot to pre-select a target, fly the aircraft, release the weapons, and leave the target area automatically without visual references.[12]

A number of A-6As were converted to perform specialized roles. These included 19 A-6Bs for "Iron Hand" missions to suppress enemy anti-aircraft and surface-to-air missile air defenses. Many of the standard attack systems were removed and replaced with special equipment to detect and track enemy radar sites and to guide AGM-45 Shrike and AGM-78 Standard anti-radiation missiles. Twelve conversions for the A-6C were for night attack missions against the Ho Chi Minh Trail in Vietnam. They were equipped with a Trails/Roads Interdiction Multi-sensor (TRIM) pod in the fuselage for Forward Looking Infrared (FLIR) and low-light television cameras, as well as a "Black Crow" detection system. (The Black Crow system was supposed to be able to home in on the radio signals generated by a truck's ignition system.)

The KA-6D was developed in the early 1970s for use as tanker aircraft, providing aerial refueling support to other strike aircraft. (78 A-6As and 12 A-6Es were ultimately converted.) The DIANE system was removed and an internal refueling system was added, sometimes supplemented by a D-704 "buddy store" refueling pod on the centerline pylon. In theory, the KA-6D could be used as a daytime visual bomber, but it never was. The standard load was four fuel tanks on the wing pylons and, because it was based on a tactical aircraft configuration, the KA-6D could provide mission tanking (the ability to keep up with strike packages and refuel them). Because they were always in short supply, they were frequently "cross-decked" from a returning carrier to an outgoing one. Due to this continual use and high numbers of catapult launches and arrested landings, many KA-6 airframes had severe G restrictions, as well as fuselage stretching. The ultimate Intruder variant, introduced in 1970, was the A-6E.

With improved avionics and the Target Recognition Attack Multi-sensor (TRAM) system (a small, gyroscopically stabilized turret, mounted under the nose of the aircraft, containing a Forward Looking Infrared [FLIR] sensor boresighted with a combination laser designator/range finder, and a new computer) the A-6E was one of the most capable strike aircraft in existence.[13] The A-6 would continue to serve until 1997.[14]

Prowler

At the same time that the original A-6 was being developed, the Navy had expressed interest in developing an electronic countermeasures (ECM) version, which Grumman developed under the designation A2F-1Q and later designated EA-6A. The EA-6A was a direct conversion of the "straight" A-6A with two seats and ECM gear replacing other equipment. The primary visual difference was the prominent fairing on the top of the tailfin for the ALQ-85 signal surveillance/receiver system. Since the wingtip airbrakes were deleted in order to mount hoop antennas under the wingtips, the EA-6A retained functional fuselage airbrakes. The forward fuselage was also stretched by about eight inches to allow the ECM gear to fit. The "Electric Intruder" could carry radio jammers, chaff pods, or fuel tanks on its underwing pylons.[15] The ECM systems were operated by an electronics countermeasures officer (ECMO) who doubled as navigator. (The ECMO was also known as a "crow" or "raven" because of a long standing identification of these birds with the electronic warfare community.)[16] Most EA-6As were retired from service with the Marines in the 1970s with the last few being retired in the 1980s. The EA-6A was essentially an interim warplane until the more-advanced EA-6B could be designed and built. Development of the much more advanced and substantially redesigned EA-6B began in 1966 as a replacement for EKA-3B Skywarrior. The forward fuselage was lengthened to create a four-seat cockpit (for the pilot and three ECMOs), and an antenna fairing was added to the tip of the vertical stabilizer similar to that on the EA-6A. Other visual differences include the gold coating embedded in the canopies (to protect against electromagnetic interference) and the tilt of the refueling probe to the right (to contain an antenna near its base). The Prowler, as the new aircraft was named, first flew in 1968 and entered service in July 1971. The first EA-6B deployment was made by a four-plane detachment on board the *America* for her third Vietnam cruise in June 1972.[17] (A total of 170 EA-6B production aircraft were manufactured from 1966 through 1991.) Due to extensive operational use, and the aircraft's age, the EA-6B has been a high-maintenance aircraft. It also has undergone more frequent equipment upgrades than any other aircraft in the Navy or Marine Corps, the first of which was the "expanded capability" (EXCAP) beginning in 1973, followed by the "improved capability" (ICAP) in 1976 and ICAP II in 1980. With the ICAP II upgrade, the EA-6B could fire both the Shrike and AGM-88 HARM (High-speed Anti-Radiation Missile) anti-radar weapons, allowing it to attack enemy radar sites and surface-to-air missile launchers on its own. (The HARM entered service in 1985 as a replacement for both the Shrike and Standard ARM.) The EA-6B is capable of gathering electronic signals intelligence. With the retirement of the EF-111 Raven in 1995, the Prowler became the only dedicated jamming aircraft in the U.S. military. Most of the Prowlers in service are the ICAP II version, but there is an ICAP III upgrade version to provide improved radar jamming and deception, as well as better threat location. (The Prowler began to be replaced by the Boeing EA-18G Growler, an ECM version of the two-seat F/A-18F Super Hornet, late in 2009.)

Hawkeye

As the threat from high-speed aircraft and long-range, air-launched guided missiles grew, the need to replace the WF-2 (E-1B) Tracer as the carrier-based Airborne Early Warning (AEW) aircraft became evident and Grumman was selected in 1957 to develop the first carrier AEW aircraft designed "from the wheels up" to incorporate the latest electronic and computer equipment for air warfare in the era of Mach 2 threats. The W2F Hawkeye was a twin turboprop aircraft with high mounted wings, but its most distinguishing feature was a 24-foot, disc-shaped "rotodome" that housed an antenna that rotated six times per minute in flight.

Because of the initial Navy requirement to be capable of operating from modernized *Essex*-class carriers, Grumman faced many engineering challenges as various height, weight, and length restrictions had to be factored into the design, with the result that some handling characteristics were adversely affected. The E-2A was introduced in January 1964 and it entered squadron service in April 1964 with VAW-11. The first deployment was on board the *Kitty Hawk* in 1965. By then problems led to the cancellation of production after 59 aircraft were built. At one point, reliability was so bad the entire fleet of aircraft was grounded. After congressional hearings into the problems, the Navy and Grumman launched a concerted effort to improve the design. The result was the E-2B with better computer electronics and larger outer-tail fins. This was the version that finally showed the promise of the basic design, but the E-2B, though it was a great improvement over the unreliable E-2A, was really still an interim design until the E-2C became operational in 1973. Because it was the successor to the "Willy Fudd" (Tracer), the E-2 was first known as the "Super Willy Fudd," but is now known more commonly as the "Hummer" for the distinctive sound made by its turboprop engines.[18]

Greyhound

Just as the earlier C-1 Trader had been developed from the S-2 Tracker for carrier on-board delivery (COD), the E-2 airframe became the basis for the C-2 Greyhound COD aircraft. Sharing the same wings and engines as the E-2 but with widened fuselage and a rear loading ramp, the Greyhound can carry up to 10,000 pounds of cargo, or passengers, or a combination of both. The first prototypes flew in 1964 and production began in 1965. The original C-2A aircraft were overhauled to extend their operational life in 1973, and in 1984, the Navy ordered new C-2A aircraft to replace older airframes.[19]

Corsair II

The Ling-Temco-Vought A-7 Corsair II, developed in the early 1960s as a replacement for the A-4 Skyhawk, was the first carrier jet aircraft designed for limited wars. (Although a nuclear attack capability was retained, the Navy had always believed in attack carriers as "general purpose forces" and not solely "strategic offensive forces.") In order to save development time, the Navy had asked for a design based on existing aircraft. LTV was selected in 1963 with a design that was derived from the successful F-8 Crusader.[20] Development was rapid. The first aircraft flew in 1965 and initial carrier qualifications were carried out in November 1966 on board the *America*.[21] The A-7A entered squadron service with the Argonauts of VA-147 late in 1966 and first deployed to Vietnam in December 1967 on board the *Ranger*. The A-7 was a subsonic version of the Crusader with a shorter, broader fuselage and wing that had a longer span, but without the variable incidence feature of the F-8. To achieve the required range, the A-7 was powered by a non-afterburning Pratt & Whitney TF30 turbofan engine. (Turbofans achieve greater efficiency by moving a larger mass of air at a lower velocity.) Although officially named the Corsair II to honor the famous Vought F4U, the A-7 was known as the "SLUFF" (for Short Little Ugly Fat Fellow—the polite form). The first model, the A-7A, was underpowered and Corsair pilots used to say "it may not be fast, but it sure is slow." Overall, however, the A-7 had a greater combat radius and carried a 60 percent greater payload than the A-4 on its six underwing pylons. It also had one 20mm HK12 cannon on each side of the nose as well as "cheek" fuselage points for Sidewinder air-to-air missiles. The A-7 featured leading-edge avionics compared to its contemporaries, with an integrated weapons computer, data link capabilities, projected map displays slaved to the inertial navigation system (INS) and a head-up display (HUD) that allowed the pilot to keep his focus outside the cockpit during operations. The A-7A was followed in combat by the improved A-7B in 1969 with more power and improved flaps.[22] The ultimate Navy Corsair, the A-7E, entered service in 1971 and included many of the features of the Air Force A-7D, such as a more powerful Allison TF41 turbofan engine, a 20mm Vulcan gatling gun and improved electronics. The A-7E deployed to Southeast Asia in May 1970 with the VA-146 Blue Diamonds and the VA-147 Argonauts on the *America*. (A-7s were also used in the mining of Haiphong Harbor, which contributed toward ending the Vietnam conflict.) Several upgrades to the A-7E, including a Forward Looking Infrared (FLIR) pod, were made over the years and the A-7E served into the mid-1990s as attack squadrons transitioned to the McDonnell Douglas F/A-18 Hornet.

7 New Developments

Even as new carrier aircraft were entering service in the early 1960s, the carriers themselves were being modified for increased military effectiveness and to improve their safety and efficiency of operation. Some changes were required by new aircraft developments, others involved entirely new concepts, while still other changes offered improvements to existing systems as new technologies emerged. The Navy would also experiment with ideas aimed at increasing operational flexibility or capability, but, as always, changes offered by improvement would be balanced against the cost of implementing them and not all experiments would be adopted.

Nose-Tow Launches

The method used to catapult aircraft from a carrier had essentially not changed since World War II. The existing bridle hookup process was labor-intensive, time-consuming, and dangerous. In 1962 the Navy started using a new system to launch aircraft. A tow bar was attached to the nose wheel of carrier aircraft so equipped. (The Grumman E-2 was the first carrier aircraft with a nose-tow system.) When preparing for launch, the bar pivoted downward to engage the catapult shuttle directly. The new system was faster, more efficient, and, as the catapult tow bar was attached directly to the aircraft nose wheel, additional equipment was not required. The first shipboard test of the new nose-tow gear was on 19 December 1962, when an E-2A was catapulted off the *Enterprise*, followed minutes later by an A-6A Intruder.[1] As newer aircraft, such as the A-7, entered service the need for bridle launches diminished, but bridle catchers continued to be a feature of Navy carriers for many years. Eventually, as older aircraft were phased out, the bridle catchers protruding from the edge of the carrier decks were removed.

Automatic Carrier Landing System

In 1962 the first SPN-10 All-Weather Carrier Landing System (AWCLS) production systems were installed on the *Midway* and *Independence* and certified for operational use in 1963. Although these systems were installed on the Navy's aircraft carriers over the next few years, their reliability was low, leading the Navy to improve them by replacing the old vacuum tube equipment with digital computers and solid state electronics. The new system became the SPN-42A. (While the SPN-42A was being developed, changes were made to the SPN-10 to improve reliability, but the automatic touchdown capability was eliminated. The system could still control aircraft to carrier approach minimums, at which point the pilots would land the aircraft manually.) In 1968, operational evaluation (OPEVAL) tests were conducted on the *Saratoga* and the system was approved for use. Over the next ten years new carriers were equipped with the SPN-42A as they were commissioned and SPN-10 systems on existing carriers were updated to SPN-42A. The SPN-42A in turn was replaced by the SPN-46, which featured fewer modules and greater reliability. The first system was installed on the *John F. Kennedy* in 1985 and OPEVAL trials were conducted in 1986 and 1987 with F-14 Tomcats. In 1987 the Navy approved full automatic control from an aircraft being acquired at ten nautical miles out, all the way to touchdown on the flight deck.[2]

The Automatic Carrier Landing System (ACLS) is not a single "box," but a combination of components that are installed on the carrier that interact with other systems on the aircraft. Once engaged, data links from the carrier to the aircraft provide roll commands to intercept and lock onto the landing pattern and pitch commands to establish the proper glide path. The Automatic Flight Control System (AFCS or autopilot) on the aircraft warns the pilot if the automatic carrier landing mode becomes uncoupled or is degraded. The Instrument Landing System (ILS) radar on the carrier transmits the glide path information to the aircraft; one antenna transmits azimuth information, and the other transmits elevation information. The radar uses a conical scan to compare the aircraft position with the desired glide path, and digital computers compute needed corrections. The aircraft is also equipped with a radar beacon to provide precise positioning information as well as automatic throttle controls and attitude indicators. The pilot can select one of three modes of operation. Mode I is a fully automatic approach from entry point to touchdown on the flight deck. In Mode II the pilot observes the cockpit displays and controls the aircraft manually. In Mode III the pilot controls the aircraft manually all the way with verbal guidance provided by the Carrier Air Traffic Control Center (CATCC) on board the carrier. (This is the normal mode for clear weather operations as described in chapter 2.)

Under Mode I, aircraft are held at the marshaling point, typically about 180 degrees from the ship's Base Recovery Course (BRC), at a unique distance and altitude, with a holding pattern in a left-hand racetrack pattern. Aircraft departing marshal are normally separated at one minute intervals with adjustments directed by the CATCC, if needed, to arrive at the radar acquisition "window." This is the approach phase. The flight from the radar window to touchdown is the descent phase.[3]

Pilot Landing Aid Television

In December 1961 the first operational Pilot Landing Aid Television (PLAT) system was installed on the *Coral Sea*. The PLAT system included a camera that pointed back along the glide slope to record every landing on high-resolution videotape. It was not only useful for pilots to analyze their landing technique, but provided a record that could be used in the analysis of landing accidents. By early 1963 all attack carriers were equipped with the PLAT system.[4]

CV Concept

As the older *Essex*-class antisubmarine carriers were coming to the end of their service lives in the early 1970s, the need for antisubmarine protection had not diminished. (The last of the *Essex*-class carriers would be decommissioned by 1976, leaving only the *Lexington* to serve until 1991 as a training carrier.) The Navy's response, supported by Admiral Elmo C. Zumwalt, who had become Chief of Naval Operations in 1970, was to return to the concept of general purpose aircraft carriers, which had been the case until 1952 when the "CVA" attack carrier designation came into use. From December 1969 to June 1970, the *Forrestal* had deployed to the Mediterranean with eight SH-3D Sea King ASW helicopters added to her air wing and the *Independence* also deployed to the Med with another eight SH-3s from July to September of that same year. These experiments were evaluated as being effective for providing close-in ASW defense for the CVA task group. The *Saratoga* was the first attack carrier to officially become a multi-purpose "CV" and began her transition during her June through October 1971 Mediterranean deployment with VS-28, equipped with the S-2 Tracker, and HS-7 with the SH-3 as part of Air Wing 3, although she was not officially redesignated until 30 June 1972.[5] During her first and only deployment to Vietnam from April 1972 to February 1973, however, only HS-7 was embarked. On her next deployment to the Med in 1974, VS-24 was embarked. The next carriers to be redesignated were the *Independence* on 28 February 1973 and the *Kitty Hawk* and *Kennedy* on 29 April 1973. On 1 July 1975, all U.S. attack carriers were redesignated as "CV" from "CVA" (or "CVN" from "CVAN").[6]

The Grumman S-2 Tracker was the Navy's first antisubmarine warfare aircraft to combine both hunter and killer roles in a single aircraft, entering squadron service in 1954 on *Essex*-class CVS carriers. It was a conventional, high-wing, twin piston engine design with tricycle landing gear. Its original designation of S2F led to the nickname of "stoof" before it was redesignated in 1962. It was replaced by the Lockheed S-3 Viking in 1976, but continued to serve in a number of navies around the world.

The Sikorsky SH-3 Sea King was the first ASW helicopter to use turbo-shaft engines, as well as the world's first amphibious helicopter. Introduced in 1961, it served with the Navy until 2006, but remains in service in many countries around the world. The Sea King was eventually replaced by the SH-60 Seahawk.

Apart from the changes in the composition of the carrier air wings, an ASW Operational Center (ASWOC) was installed to service the computer systems of the anticipated S-3 Viking aircraft and provide other support to S-3 operations.[7] (The

ASWOCs later became Tactical Support Centers [TSC] when the VS antisubmarine squadrons expanded their operations to include sea control missions. The S-3 and SH-60 are covered in chapter 11.)

New Weapons

Sea Sparrow

Since the late 1950s the threat from high-speed jet aircraft approaching ships at low altitudes presented a serious threat to naval forces and the introduction of sea-skimming anti-ship missiles (ASM) dramatically increased the threat. With the introduction of the *Kitty Hawk* class, the Navy acknowledged that guns were no longer effective and introduced the Terrier missile system for carrier defense. When the Navy later decided to put its more capable air defense missile systems on board the escort ships in the carrier battle groups, there was still a need for a self-defense system for the carriers that could respond quickly to incoming missiles. After attempts to adapt the Army's Mauler system into a Point Defense Missile System for ships ran into development problems, the Navy decided to adapt the AIM-7E Sparrow to fill the gap. As the Basic Point Defense Missile System (BPDMS) this system was relatively simple and straightforward: the launcher was adapted from the existing antisubmarine rocket (ASROC) units used on destroyers and the radar was a manually trainable unit with continuous wave illuminators that looked like two large searchlights. The operator could be cued by voice from the ship's search radar operators and the target did not have to be centered in the beam to be tracked effectively. Since the launcher automatically followed the motions of the illuminator, it would immediately see the return from a target when the missile was fired. Originally installed on smaller escort ships, the Sea Sparrow was installed on carriers in place of their original gun or Terrier missile armament. (When the *Forrestal* was repaired following her fire in 1967, her guns were removed and a launcher was added to a sponson on the starboard side forward. Two launchers were fitted to the *Independence* in 1973, two on the *Saratoga* in 1974, and two on the *Forrestal* in 1976. The *Ranger* was fitted with a system but retained her after two 5" guns until 1977.[8] *Saratoga*, *Forrestal*, *Independence*, *Kitty Hawk*, and *Constellation* received the improved systems during their Service Life Extension Program [SLEP] overhauls.) In cooperation with NATO, the Sea Sparrow has been continually updated through the Improved Basic Point Defense Missile System (IBPDMS) and the Evolved Sea Sparrow Missile (ESSM).

Close-In Weapons System

The Phalanx Close-In Weapons System (CIWS) (usually pronounced "Sea Whiz") is an anti-ship missile defense system designed by the General Dynamics Pomona Division (now a part of Raytheon). With a radar-guided six-barrel 20mm gatling gun mounted on a swiveling base, it is used on every class of Navy surface combat ship, as well as the navies of 16 allied nations. (Because of its distinctive barrel-shaped radome and its automated operation, CIWS units are sometimes called "R2-D2" after the famous droid from *Star Wars*, or as "Daleks" in the Royal Navy after the aliens from *Doctor Who*.) The CIWS was developed from the M61 Vulcan gatling cannon, used on nearly all U.S. fighter aircraft since the 1960s, and was first operational on the *America* in 1980. With a rate of fire of 3,000 to 4,500 rounds per minute and capable of fast elevation and traverse speeds to track incoming targets, CIWS is an entirely self-contained unit (the gun, automated fire control system, and all other major components included on the mount), enabling it to search, detect, track, and engage targets automatically. As both the threat and computer technology have evolved, CIWS has been upgraded to improve its capabilities against surface targets as well as supersonic anti-ship missiles and now includes infrared sensors and links to the Rolling Airframe Missile (RAM) system.

Rolling Airframe Missile (RAM)

The RIM-116 Rolling Airframe Missile (RAM) is a small, lightweight, infrared homing surface-to-air missile originally intended as a point-defense weapon against anti-ship cruise missiles. (A RAM rolls around its longitudinal axis to stabilize its flight path, not unlike a rifle bullet.) The RAM system was first operational in 1992 and is used by the American, German, South Korean, Greek, Turkish, Saudi, and Egyptian navies. On American ships it is integrated with the ship's own defense systems and has been installed on board all the nuclear-powered carriers. Of the conventionally powered carriers, only the *Kitty Hawk* and *Kennedy* were each equipped with two RAM systems.[9]

Naval Distillate Fuel

The Navy had used Navy Special Fuel Oil (NSFO), a refined form of black oil, to power its warships since before World War II. On aircraft carriers, there had always been a tradeoff between storage of fuel for ship propulsion and that for aircraft. By the 1960s the Navy was using three types of fuel: NSFO for generating steam, diesel fuel for smaller vessels and auxiliary power, and JP-5. In 1967 the Navy began to investigate the possibility of developing a single fuel for all uses. Since JP-5 cost twice as much as NFSO at the time, and there was uncertainty that enough JP-5 would be available to meet the demand for all uses, it was decided that JP-5 would not become the multipurpose Navy fuel. Although the search for one common fuel was ultimately unsuccessful, the Navy did develop a single distillate fuel for ship propulsion, Naval Distillate Fuel (NDF). NDF has been standardized with our allies as NATO F-76. JP-5 can be used as an alternate fuel if necessary. In the 1970s Navy ships were converted to use NDF as they underwent modernization and upkeep.[10]

Service Life Extension Program

By the late 1970s the *Forrestal*-class ships had been in service for a number of years and were beginning to show the effects of continuous operations. As ships built in the 1950s and 1960s, they would reach the end of their expected 30-year service lives between the mid-1980s into the late 1990s. Since the replacement *Nimitz*-class carriers in the pipeline were not going to be available anytime soon, the Navy decided to rehabilitate and update the conventionally powered carriers in a reconstruction effort known as SLEP. The SLEP program originally applied to the four ships of the *Forrestal* class, but was later expanded to cover the later ships of the *Kitty Hawk* class as well. SLEP went beyond the usual overhaul because it was not only intended to add another 15 years of service life, but also to include modernization for enhanced operational capabilities. First, there would be an extensive overhaul and refurbishment of the basic hull, machinery and electrical systems, as well major repairs to the inner bottom and bilge tanks, the propulsion plant, and piping systems. Improvements to combat systems included new SPS-48C long-range three dimensional radars; SPS-49 long-range two dimensional radars; Point Defense Missile Systems (NATO Sea Sparrow) with SPS-65 radars for air search and target acquisition; WLR-8 electronic countermeasure (ECM) receivers; Vulcan-Phalanx 20mm Close-In Weapons Systems (CIWS); Naval Tactical Data Systems (NTDS); and ASW Tactical Operations Centers (ASWOC). Other improvements included increasing weapons elevator capacity, replacing the existing arresting gear with the latest Mark 7 Mod. 3 version and crew habitability improvements.[11] The Philadelphia Naval Shipyard performed all the work. The *Saratoga* was the first to undergo SLEP. She entered the navy yard on 1 October 1980 and emerged 28 months later. She was followed by the *Forrestal* from January 1983 to May 1985, the *Independence* from February 1985 to June 1988, the *Kitty Hawk* from July 1987 to March 1991, and the *Constellation* from February 1990 to March 1993. (The *Ranger* did not receive a SLEP. Unlike the other units of her class, she retained her forward gun sponsons and her after two 5" mounts until 1977.)

Odd Experiments

Hercules

One of the oddest experiments involving a non-carrier aircraft operating from an aircraft carrier occurred in the fall of 1963 when a four-engine Lockheed KC-130F piloted by Lieutenant James H. Flatley III set the record for the largest and heaviest aircraft ever to operate from an aircraft carrier. On 30 October Flatley made 44 approaches to the *Forrestal* as she steamed out in the Atlantic. Of these approaches 16 were touch-and-go landings and the rest were intentional wave-offs made to test the Hercules' responsiveness during the approach. On 8, 21, and 22 November, Flatley and his crew made 21 full-stop landings on and takeoffs from the ship. The tests were conducted at a number of different weights to determine if the Hercules could serve as a "Super COD" (Carrier Onboard Delivery) aircraft. Although the tests were highly successful, the idea was considered too risky for routine COD operations and the C-2 Greyhound was developed as a dedicated COD aircraft instead. (Flatley was awarded the Distinguished Flying Cross for his role in the tests, and the Hercules used is now the National Museum of Naval Aviation at Pensacola, Florida. The idea of using the C-130 was looked at again in 2004–2005 as the Navy developed its "sea basing" concept for large prepositioned ships, but was still not regarded as a practical solution.)[12]

Whale Tale

During the early years of the Cold War, the famed Lockheed "Skunk Works" developed a super-secret, high-altitude reconnaissance plane, the U-2, to spy on the Soviet Union. These flights, flown for the Central Intelligence Agency (CIA), ended abruptly in May 1960 when a U-2 flown by Francis Gary Powers was shot down by a Soviet SA-2 missile. This did not end the career of the U-2, however; the spy planes continued to fly missions in other parts of the world. Even with a range of 3,000 miles, there were areas of interest to the U.S. government that could not be flown from secure land bases. So, in mid-1963, the CIA launched Project Whale Tale to adapt the U-2 for carrier operation. Since the U-2 was essentially a jet-powered glider, it could take off unassisted from a carrier deck and its slow approach speed made carrier landings relatively easy when the arresting gear was set at its lowest setting. On an August night in 1963, a U-2C was craned on board the *Kitty Hawk* at the North Island Naval Air Station in San Diego. The next day, as the Hawk steamed off the California coast, a Lockheed test pilot took off and made a number of practice approaches. As the aircraft attempted a landing it bounced and hit a wingtip before just managing to get airborne again at the end of the flight deck. The tests were deemed a success, however, and three U-2As were modified to become U-2Gs with beefed up landing gear, an arresting hook, and wing spoilers to cancel aerodynamic lift on landing. The first carrier tests of these aircraft took place in March 1964, probably on the *Kitty Hawk*. The only operational mission, Operation Seeker, was launched from the *Ranger* in May 1964 to monitor French nuclear tests at Muroa Atoll in French Polynesia. The U-2R variant, larger and heavier with folding wing panels to allow the use of carrier elevators, was tested on the *America* off Virginia in November 1969. In spite of the technical success of the program, the CIA decided that aircraft carriers were too conspicuous, and too expensive, to continue since a lone U-2 could be deployed overnight to a remote airfield for operations.[13]

Quiet Short-Haul Research Aircraft

Although it had experimented with Ling-Temco-Vought XC-142A Vertical/Short Take-Off and Landing (VSTOL) transport on the *Essex*-class carrier *Bennington* in 1966, the Navy occasionally evaluated other large four-engine cargo planes on board carriers. In 1980 the Navy evaluated the National Aeronuatics and Space Administration's Quiet Short-Haul Research Aircraft (QSRA) on the *Kitty Hawk*. (The QSRA was developed from the de Havilland Canada DHC-5 Buffalo with an experimental wing and four split-flow turbofan engines to test Short Take-Off and Landing [STOL] concepts.) The tests on board the *Kitty Hawk* were conducted without using either the catapults or arresting gear.[14]

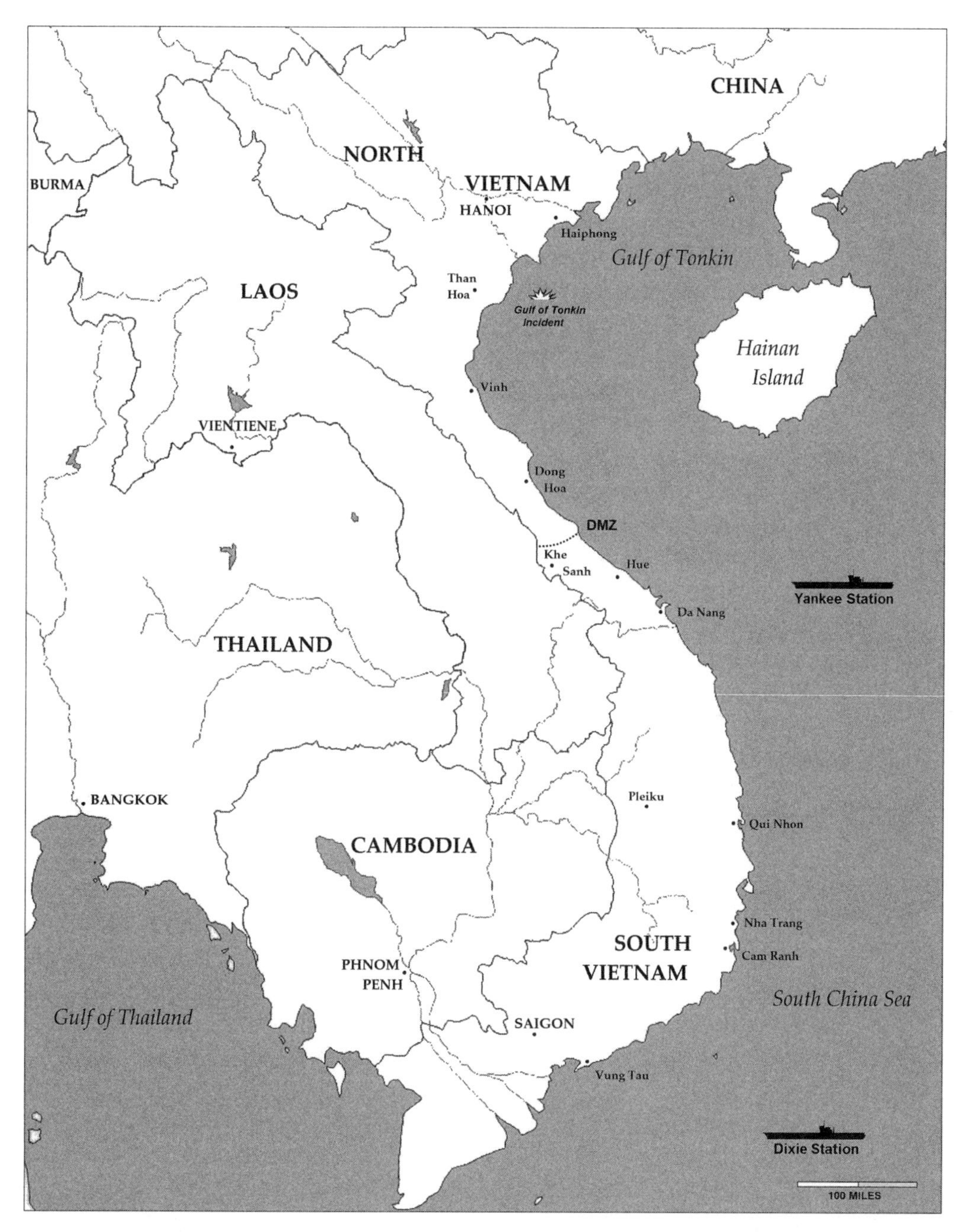

Map 3. Southeast Asia

Map adapted by the author from http://d-maps.com/carte.php?num_car=14666&lang=en.

8 Vietnam

Throughout the late 1950s and early 1960s U.S. carriers maintained a presence in the South China Sea as the turbulence in Vietnam grew following the French withdrawal from Indochina in 1954. The whole of Southeast Asia seemed on the verge of a communist takeover as guerrilla warfare continued in what had become South Vietnam and as the conflict spread to neighboring Cambodia and Laos. American involvement began with supplying equipment and advisers. After Lyndon B. Johnson assumed the presidency, contingency planning began for possible attacks against North Vietnam. By 1964 the ships of the Seventh Fleet were operating off the Indochina Coast on a routine basis, gathering intelligence, supporting anti-communist activities and launching carrier-based reconnaissance flights. In May, the Seventh Fleet initiated a standing carrier presence at Yankee Station in the Gulf of Tonkin.

In April the *Kitty Hawk* had been sent to the South China Sea to provide a stabilizing presence as communist Pathet Lao guerillas threatened the Laotian government. The Hawk started flying low-level reconnaissance missions over Laos in May, monitoring the communist infiltration into the area from North Vietnam, over part of what eventually became known as the Ho Chi Minh Trail. On 6 June, while on a reconnaissance mission east of the Plaines des Jarres, machine gun and 37mm anti-aircraft fire brought down an RF-8 Crusader from the Hawk's light photographic squadron VFP-63 detachment. The pilot, Lieutenant Charles F. Klussman, was injured during the ejection and was captured by the Pathet Lao. After enduring 86 days of brutal captivity, he escaped and made his way to a pickup site in Laos where he was rescued by an Air America light aircraft. Klussman had become the first naval aviator to fall into enemy hands during the Vietnam war and would later be awarded the Distinguished Flying Cross (DFC) for his exploit. (Air America was owned and operated by the Central Intelligence Agency and operated extensively throughout Southeast Asia during America's involvement.)[1]

The Gulf of Tonkin

On the afternoon of 2 August 1964, as she operated off the coast of North Vietnam, the destroyer *Maddox's* radar detected three unidentified high-speed craft. The North Vietnamese torpedo boats ignored the warning shot from the *Maddox* and two of them fired torpedoes. The *Maddox* evaded the torpedoes and returned the fire, possibly hitting the third North Vietnamese boat. Meanwhile, a training flight of four F-8 Crusaders from the *Ticonderoga* was vectored to the area. After contacting the *Maddox*, the Crusaders were ordered to attack the torpedo boats as they headed north. Armed with 20mm cannon and unguided Zuni rockets, they made several strafing runs over the craft, sinking the third torpedo boat. The whole affair lasted about three and one half hours. President Johnson ordered the destroyer *Turner Joy* to join the *Maddox* and the *Constellation* was routed from Hong Kong to the Gulf of Tonkin. Meanwhile the *Ticonderoga* maintained daylight patrols and the destroyers retired to about a hundred miles off shore during the night to reduce the danger of torpedo boat attack.

On the night of 4 August, the *Maddox* picked up five high-speed radar contacts, identified as North Vietnamese torpedo boats about to make nighttime runs. In bad weather, the two forces exchanged gunfire and torpedoes from the boats narrowly missed the destroyers. *Ticonderoga* launched two A-1 Skyraiders to provide air cover, but by then the destroyers had lost radar contact.

Retaliation was not long in coming. Even as President Johnson went on national television to speak about the two unprovoked attacks on American warships in international waters and to announce the actions he would take, aircraft from the *Ticonderoga* and *Constellation* were heading for four major North Vietnamese torpedo boat bases. In strikes that lasted more than four hours, Crusaders, Skyhawks, and Skyraiders bombed and rocketed the four bases, damaging all the facilities and destroying or damaging an estimated 25 torpedo boats and more than half of the North Vietnamese air force. *Constellation* lost an A-1 Skyraider, whose pilot, Lieutenant (jg) Richard A. Sather, became the first Navy pilot to be killed in Vietnam and an A-4 Skyhawk, flown by Lieutenant (jg) Everett Alvarez, who became the first Navy prisoner of war (POW).[2] With the passage of the Gulf of Tonkin Resolution by Congress a few days later, Task Force 77 stood ready to launch further retaliatory attacks against the North. The *Ranger* and the *Kearsarge* joined Task Force 77 in the Gulf of Tonkin, giving the task force three attack carriers and one antisubmarine carrier. The communists shifted a few fighter units from China into North Vietnam and the Americans brought Air Force units into South Vietnam, along with support personnel and equipment. The buildup had begun. For the next six months, American carriers continued to patrol off Vietnam.

Although there were occasional reports of communist terrorist activities in Saigon and the countryside, things remained quiet until February 1965. Dissatisfaction with General Nguyen Khanh's government led to street demonstrations and Viet Cong attacks, and a U.S. advisory team, led by National Security Advisor McGeorge Bundy, was sent to observe the South Vietnamese situation and make recommendations. On

7 February, shortly after the team was to leave, the Viet Cong attacked the American compound in Pleiku in the Central Highlands, killing nine Americans and wounding a hundred. To the American officials in South Vietnam, there seemed little choice but to retaliate and President Johnson authorized a Navy strike, coordinated with South Vietnamese aircraft, against the military barracks and staging area at Dong Hoi, north of the 17th parallel. The strike, named Flaming Dart I, involved the *Coral Sea*, *Ranger*, and *Hancock*, but weather in the South China Sea caused a delay. By noon, however, the carriers were ordered to launch their aircraft. *Coral Sea* and *Hancock* were to hit the barracks at Dong Hoi while the *Ranger* would hit barracks 15 miles inland at Vit Thu Lu. The *Ranger's* aircraft could not hit their target because of the weather, but the strikes against Dong Hoi destroyed much of the facility. The communists responded with an attack on a hotel in Qui Nhon used as American enlisted quarters. Twenty-three American soldiers were killed and many wounded. Flaming Dart II began the next day, 11 February, with strikes against the Chanh Hoa barracks 35 miles north of the demilitarized zone (DMZ). Flying in the northeast monsoon, this strike force faced the same rain and fog that had hampered the 7 February strike. The Chanh Hoa strike had limited success and the attacking aircraft faced heavy antiaircraft artillery (AAA) fire. Three aircraft were shot down and others damaged. These strikes were the first of the so-called Alpha Strikes, which involved using all available aircraft from a carrier's air wing, from fighters to tankers. The communists, unimpressed by the American carrier strikes, continued their periodic attacks against American installations in South Vietnam. The monsoon weather continued to plague carrier operations for the rest of the war.

Rolling Thunder

Because merely responding to communist attacks did not prove effective, President Johnson gradually allowed a program of interdiction air strikes. The Air Force and Navy were tasked to bomb targets successively further north of the DMZ. As the bomb line of these Rolling Thunder operations approached the capital of Hanoi, the North Vietnamese were expected to sue for peace. Severe restrictions imposed on the operational commanders hamstrung the operation from the start—no pre-strike photography was permitted, reconnaissance aircraft had to accompany the strike or fly in immediately afterward, no follow-up strikes were authorized, unexpended ordnance could not be dropped on targets of opportunity, and enemy aircraft had to be positively identified before engaging in air-to-air combat. Specific targets were assigned to the Navy and Air Force, and later these were organized into "route packages" to relieve some of the confusion in planning missions. Two geographic stations were picked in the Gulf of Tonkin—Yankee Station in the north and Dixie Station in the south. Yankee Station served as the center of carrier operations against North Vietnam while Dixie Station became a "warm up" area where newly arrived carrier air wings could gain experience while supporting ground operations in South Vietnam before rotating to the "big leagues" up north. The first Navy Rolling Thunder strikes were launched by the *Coral Sea* and *Hancock* against supply buildings at Phu Van and

Vinh Son on 18 March 1965. All the aircraft returned with light damage, but successive Rolling Thunder strikes brought increased losses as the attacks moved north to within 70 miles of Hanoi. Although Rolling Thunder was somewhat effective, the communists continued the war by moving and resupplying their units at night over the Ho Chi Minh Trail from North Vietnam into Laos.

Iron Hand

As the tempo of air strikes increased, the North Vietnamese responded with stronger air defenses and in April, the first surface-to-air missile (SAM) site was photographed 15 miles southeast of Hanoi. By July several SAM sites were detected, but it was not until August, after several aircraft were lost, that permission was given to attack the SAM sites. These missions became known as Iron Hand in the Navy and Wild Weasel in the Air Force. Navy Iron Hand operations began in August 1966, but the first actual strike against a SAM site was on 17 October, when four A-4s from the *Independence*, with an A-6 pathfinder, found a site near Kep airfield north of Hanoi and destroyed it. A common tactic was for A-4s or A-6s armed with Shrike anti-radiation missiles to fly at a low-level "above deck" altitude where they could be detected by the SAM radars. When the Iron Hand aircraft's electronic warning gear indicated that the SAM site had "locked on" the strike aircraft would dive for the "hard deck" altitude where the SAM radar could not track it, while changing course and then pitching up 15 degrees to launch the Shrike into the SAM radar's detection cone. (The AGM-45 Shrike anti-radiation missile had been developed by the Naval Weapons Center at China Lake in 1963 by the simple expedient of putting a radar seeker head on the body of an AIM-7 Sparrow air-to-air missile.) Although its range was shorter than that of the SA-2 missiles it was used against, it was a great improvement over the previous method of trying to take out SAM sites with rockets and bombs. Since the Shrike was not a true "fire and forget" missile, it could not "remember" where a site was if the radar signal was lost. The North Vietnamese eventually began to get wise to this and started using their radars intermittently or shutting them off completely if they felt threatened. Later, the Shrike was supplemented by the Standard ARM, often referred to as a "starm." (The AGM-78 Standard ARM was developed by General Dynamics in the late 1960s because of the limitations of the Shrike, which had a small warhead, limited range, and a poor guidance system. Modifying an "off the shelf" RIM-66 SM-1 surface-to-air missile to reduce development costs, it was carried by modified Intruder, the A-6B. Later versions had improved broadband seekers and memory circuits to guide the missile even if the enemy radar had shut down.)

The SA-2 Guideline missile (NATO code name) was Soviet-designed, high-altitude, command guided missile first deployed in 1957. (A longer-range and higher-altitude version had brought down the U-2 flown by Francis Gary Powers over the Soviet Union in May 1960. The system was also deployed in Cuba during the Cuban missile crisis.) The North Vietnamese used the SA-2 extensively during the Vietnam war to defend Hanoi and Haiphong. It was also locally produced by the People's Republic of China. The missile itself was only one part of an overall air defense system. A Spoon Rest early warning radar usually detected incoming aircraft and handed off to the Fan Song guidance radar. This radar had the ability to track targets while continuing to scan for others. Initially, the North Vietnamese followed Soviet doctrine in laying out their missile sites, with six launchers about 200 to 300 feet apart arranged in a hexagonal star pattern with the radars and guidance systems in the center. Another six missiles on tractor-trailers were usually stored nearby. The distinctive pattern of the access roads made the SAM sites readily identifiable from reconnaissance photography and the sites were later arranged in less obvious patterns and were better camouflaged.[3]

SAMs were not the only problem; the North Vietnamese set up early warning radar sites as fast as possible and introduced growing numbers of MiG-17 jet fighters. The MiG-17 was the follow-on improvement of the famous MiG-15 from the Korean War. Although slower than the American fighters, the MiG-17 was very maneuverable and its cannon armament could be deadly. Later, the more advanced delta-winged MiG-21 appeared over North Vietnam, posing an even greater threat. Also, as attacking aircraft were forced down to lower altitudes to get out of the performance envelope of the Soviet-designed SA-2 Guideline missiles, they faced radar-controlled antiaircraft guns. The flak, according to some veterans, was as intense as any they had experienced in World War II. The three carriers on station at the end of 1965, the *Enterprise*, *Kitty Hawk*, and *Ticonderoga*, wound up the year with one of the biggest strikes flown up to that time. One hundred aircraft hit the thermal power plant at Uong Bi on 22 December 1965, marking the first time that an industrial target, as opposed to bases and support installations, had been hit.

By the time the Christmas truce began on 24 December 1965, ten carriers had seen combat since August 1964 and many of them now rotated home—first the *Coral Sea*, then *Midway*, *Independence*, *Bon Homme Richard*, and *Oriskany*. The *Enterprise* and *Kitty Hawk* were left to carry on. The bombing halt lasted for 37 days and while the Americans waited for peace talks to begin, the North Vietnamese rebuilt their damaged bridges and facilities and added to their air defenses. Rolling Thunder had failed. Nearly 57,000 combat sorties had been flown, with over a hundred aircraft lost, and 82 men killed, captured, or missing. Forty-six were rescued. Ironically, the carriers had shown their worth by carrying the war to North Vietnam and in supporting operations in the south.

Under pressure from senior military commanders, the administration gave approval for a resumption of the bombing campaign, although many restrictions, such as avoiding foreign shipping, MiG airfields, and large industrial targets north of Hanoi, remained. Gradually, the strikes were directed more at interdicting the flow of men and supplies into the south than "punishing" the North Vietnamese. The northeast monsoon, which runs from November through April, was at its height in the South China Sea when the bombing resumed in January 1966. Although SAMs were a threat, most of the losses were due to heavy flak. Pilots had learned that the SA-2 Guideline missile could be beaten with the right tactics. When a pilot sighted a missile, which was described as looking like a flying telephone pole, coming at him, he would wait until the proper moment and then execute a hard turn across the missile's flight path. With luck, the missile seeker would break lock and eventually explode without effect. Only the pilot being shot at could decide when to pull the stick and waiting for the right moment took courage, skill, and stamina.

During the early months of 1966, shortages of pilots and aircraft began to effect operations. Aircraft production was not keeping pace with attrition and the pilot shortage was aggravated by a combination of combat losses and pilots leaving the service out of frustration over the restrictions imposed or the lure of lucrative airline jobs. Pilots flew an average of between 16 and 22 combat missions a month over the North with some going as high as 28 a month. The Defense Department eventually decided that a pilot could not fly more than two carrier deployments within a 14-month period, but even with these restrictions there were individuals who completed over 500 missions. More pilots were training, but it would be 18 months, the normal training cycle for a pilot to earn his wings, before any improvement would be felt. In March the monsoon abated slightly and the air crews welcomed the clearer weather, but better weather meant increased operations in March and April. Loss rates climbed as strikes against industrial areas of the North increased. During April North Vietnam was divided into target areas with the Navy assuming responsibility for the coastal areas, especially Haiphong. Crews could get to know the defenses of their areas and plan to avoid them as much as possible. In March the first MiG-21s had been sighted and more MiG-17s were appearing, but it was not until 12 June that the first MiG kill was scored by an F-8 pilot from the *Hancock*, and another *Hancock* pilot scored again nine days later.

At the end of June Defense Secretary Robert S. McNamara announced the beginning of a campaign directed against the North's petroleum facilities. The POL (for petroleum, oil, and lubricants) campaign, known as Rolling Thunder 50, marked a new and more dangerous phase of the air war as the oil and industrial areas in the northeast were attacked. The POL campaign continued through 1966 and into 1967. Large strikes in July and August took out major parts of the communist facilities, including transportation equipment, such as trucks, rolling stock, and barges. The *Constellation*, on her second deployment to Vietnam with Carrier Air Wing 15, scored her first MiG kill of the war on 13 July when an F-4B flown by Lieutenant William M. McGunigan and radar intercept officer Lieutenant (jg) Robert M. Fowler from the VF 161 Chargers shot down a MiG-17 fighter with a Sidewinder.[4] Strikes from the *Franklin D. Roosevelt*, *Constellation*, *Ranger*, and *Hancock* ranged from the DMZ to Haiphong. In October 1966 an F-8 from the *Intrepid* scored the Navy's first victory over a MiG-21. (The *Intrepid* had joined in the strikes during September and October.) Although crews were elated that they could finally hit the right targets, the North Vietnamese had used the time to cache much of their oil in underground bunkers and tanks. The free zone around Hanoi, in which no bombs could be dropped for fear of hitting population centers, and the restrictions on striking ports, such as Haiphong for fear of hitting foreign ships, remained. Despite American efforts, the North was determined to carry on the war.

On 26 October 1966 fire broke out on the *Oriskany* when two sailors were returning some unused flares from a strike. When one of them ignited, one of the sailors panicked and threw it into a locker. The resulting fire spread to the hangar bay, setting off other ordnance. By the time the ordeal was over 44 men had died. Fire would continue to pose a danger as the war went on.

During her second WestPac deployment to Vietnam, the *Kitty Hawk* scored her first kills on 20 December 1966, when two F-4Bs, flown by Lieutenant Hugh D. Wisely and Lieutenant (jg) David L. Jordon of the VF-114 Aardvarks and Lieutenant David A. McRae and Ensign David N. Nichols of the VF-213 Black Lions, intercepted and shot down a pair of North Vietnamese An-2 Colt single-engine propeller-driven biplane utility aircraft. Both Phantoms had been scrambled early that morning to investigate an unidentified radar contact. After tracking the Colts to about 25 miles east-northeast of Than Hoa in North Vietnam, they were brought down with Sparrow missiles. Wisely later recalled, "I saw the missile explode, and saw an explosion as the plane erupted."[5] At the beginning of 1967, air-to-air combat picked up, with the Air Force tangling with MiGs over North Vietnam. The MiG activity died down, but resumed in March. Airfields could no longer be kept off limits and on 24 April, the airfield at Kep, 37 miles northeast of Hanoi, was struck by the *Kitty Hawk* and *Bon Homme Richard*. The runway was damaged along with several MiGs on the ground. F-4s from the *Kitty Hawk* downed two MiG-17s as they took off to intercept. Two VF-114 Phantoms engaged two flights of North Vietnamese MiG-17s, one flight "low over the ground" and the other flight of three "directly ahead." The two Phantoms shot down a pair with Sidewinders, Lieutenant Hugh D. Wisely and Lieutenant (jg) David L. Jordon again scoring (this time against tougher opponents than the propeller driven An-2 Colts) and Lieutenant Commander Charles E. Southwick and Ensign James W. Laing. The F-4 flown by Southwick and Laing, however, ran out of fuel and had to ditch at sea, but both men were rescued by a Navy SAR helo.[6] Two more MiGs fell to the F-8s from the *Bon Homme Richard*. As the Air Force struck air bases in the North, April and May became periods of intense air-to-air combat as MiGs rose to challenge strike groups. (Of the six Navy kills during April and May, five were scored by F-8s from the *Bon Homme Richard*. An A-4 pilot got the other MiG by entering the landing pattern over a North Vietnamese airfield and shooting it down with an unguided Zuni rocket.)

Early in 1967, the decision was made to begin mining selected rivers as the mounting air attacks on road and rail traffic forced the communists to shift more to the rivers leading into the South. In March the new Walleye television-guided air-to-surface glide made its combat debut with A-4s from the *Bon Homme Richard*. Bridges had become a favorite target and, although they formed a major part of the Vietnamese road system, they proved to be extraordinarily difficult targets. They were not only difficult to hit and destroy, but were also heavily defended and therefore costly targets. The resourcefulness shown by the North Vietnamese in repairing damaged bridges or bypassing them added to the frustration. The emphasis on bridges was due in part to the prohibition on attacking Hanoi and Haiphong. If Haiphong, a center of supply operations, could not be hit, then it could at least be isolated by destroying approaches outside the main area. As the months wore on, attacks on bridges, roads, and canals were met with intense flak. At one point,

Haiphong apparently ran out of ammunition as striking aircraft met no SAMs and only light flak. The respite lasted two days before weather precluded strikes for the next three. When the rain cleared and operations resumed, so did the flak and missiles. Haiphong had been resupplied. On 29 July, at the height of this effort, the *Oriskany*, herself the victim of a fire the previous October, would stand by to help another victim of fire, the *Forrestal*.[7]

9 The *Forrestal* Fire

On the morning of 29 July 1967 the *Forrestal* was operating in the Gulf of Tonkin on Yankee Station on the fifth day of her first combat deployment. Over the previous four days she had launched over 150 sorties without the loss of a single aircraft. At about 1050, as the second launch of the day was being readied, a Zuni rocket fired accidentally from an F-4B Phantom parked on the starboard side at the aft end of the flight deck. It hit the drop tank of an A-4E parked across the deck on the port side, aircraft side number 405, piloted by Lieutenant Commander Fred White of VA-46. The Skyhawk erupted in a ball of flame and White died in the fire.[1] Lieutenant Commander John McCain, the pilot of Skyhawk side number 416 next to White's, was among the first to notice the flames and escaped by climbing over the nose of his aircraft and jumping off the refueling probe shortly before the explosions began.[2] The wind spread the flames quickly and the after end of the flight deck was engulfed in flame and smoke as other munitions began to explode and fuel from damaged aircraft spread the fire. Berthing spaces immediately below the flight deck became death traps for 50 men, while other crewmen were blown overboard by the explosions. Others escaped into the catwalks around the edge of the flight deck. As crewmen struggled to clear the deck of ordnance, throwing bombs and ammunition over the side, a chief petty officer leading the initial firefighting party rushed forward with a hand held fire extinguisher to beat down the flames so that the pilots might escape. As he approached, he spotted a bomb already surrounded by flames—it exploded, killing him instantly.

The ordeal of the *Forrestal* was only beginning. The firefighting crews on the flight deck rallied after the first explosion and attacked the fire again, only to disappear in the second, greater series of explosions. The decimated firefighters sought help from anyone on the flight deck, and these improvised crews once again moved into the growing inferno. A third round of detonations cleared the deck of men and firefighting gear, but within minutes more crewmen from the forward deck and below deck areas had reconstituted firefighting teams and were working their way aft. By the time the fires were brought under control 134 *Forrestal* men had died, 161 were injured, 21 aircraft were destroyed, and 41 damaged. Seventy two million dollars in damage had been done, and it would require seven months to repair the carrier. This disaster was the result of a chain of events that would leave a lasting impression on the Navy and change the way it regarded safety at sea.

Background

During the Vietnam war, several carriers normally assigned to the Atlantic Fleet were occasionally rotated to the Pacific for duty in WestPac. On 6 June 1967 *Forrestal* left her home port in Norfolk for what was to be her first deployment to Vietnam. On board was Air Wing 17 with two fighter squadrons, the VF-11 Red Rippers and VF-74 Be-Devilers, with F-4Bs; two attack squadrons, the VA-106 Gladiators and VA-46 Clansmen, flying A-4Es; reconnaissance squadron RVAH-11, the Smokin' Tigers, with RA-5C Vigilantes; KA-3Bs of the VAH-10 Vikings; and the VAW-123 Screwtops, flying E-2As. All were East Coast squadrons. After completing the inspections required for the upcoming cruise, she went on to Brazil for a show of force before sailing around the Horn of Africa to the Philippines, where she docked for a short while at Leyte Pier at the Cubi Point Naval Air Station in Subic Bay. Because of a shortage of 1,000-pound bombs, AN-M65 bombs of World War II vintage were loaded on board instead of the newer low drag Mark 82 500-pound and Mark 83 1,000-pound bombs. This would have disastrous consequences—the old bombs were filled with Composition B explosive, while the Mark 82s and Mark 83s were filled with the newer Composition H6 explosive, which was capable of withstanding higher temperatures before "cooking off." The Zuni rocket that struck Fred White's A-4 also dislodged two bombs—AN-M65s—that fell to the deck in a pool of burning JP-5 jet fuel. Because of their training, the firefighting teams on the *Forrestal* thought they had almost three minutes to reduce the temperature of the bombs to a safe level, but they did not realize the Composition B bombs were already critically close to cooking off.[3] Chief Aviation Boatswain's Mate Gerald W. Farrier, the chief who had rushed forward with the fire extinguisher, realized that a lethal explosion was imminent when one split open. He shouted for the firefighters to withdraw but the bomb exploded seconds later—only a minute and a half after the start of the fire.[4]

The Inferno

The epic struggle to control the inferno is the story of thousands of individual acts of courage as the men of the *Forrestal* struggled to save their ship and their shipmates. Nearly all of the *Forrestal* crewmen who knew how to properly fight the fire died in the initial explosions. Those crewmen who volunteered to fight the fires had not been trained on the firefighting equipment or the proper methods of using it. While one team would be spraying the fire with fire-retarding foam, the proper procedure, in another area a different team would be spraying seawater on the fire, which only served to wash away the foam. Worse, the seawater washed the flaming jet fuel through the gaping holes in the flight deck, spreading fires to the decks below, down to the berthing spaces on the 03 level under the aft end of the flight deck, and then onto the 02 and 01 levels and Hangar Bay 3. But the men of the *Forrestal* would not be alone in their efforts. A helicopter from the *Oriskany* that had been tasked to fly plane guard for *Forrestal* that morning after completing a flight to that carrier, would begin rescuing *Forrestal* crewmen who had jumped, fell, or were blown overboard and would later shuttle medical supplies to the stricken ship. *Oriskany* and *Bon Homme Richard* provided medical teams and firefighting equipment. The destroyers *Rupertus* and *George K. MacKenzie*, in what was later called an act of "magnificent seamanship," had maneuvered their ships to within 20 feet of the carrier so their fire hoses could be used effectively.[5] By 1147 the skipper of the *Forrestal*, Captain John K. Beling, could report that the flight deck fire was under control and at 1215 the ship sent word that the flight deck fire was out. All available COD (Carrier Onboard Delivery) aircraft were sent to the carriers *Oriskany* and *Bon Homme Richard* to be rigged with litters for medical evacuation. Although all the ship's vital machinery and steering equipment were operational, and the ship was never in danger of sinking, the fires in hanger bay 3 and the 01 and 02 levels still burned. At 1412 the after radio compartment was evacuated because of dense smoke and water, but the ship could report, "All fires out on 01 level, port side." Even as the compartment fires continued, progress was being made as the *Forrestal* steamed toward a rendezvous with the hospital ship *Repose*.

The commander of Task Force 77 announced at 1500 that he was sending *Forrestal* to Subic Bay, after the carrier rendezvoused with the hospital ship *Repose*. At 1705, even as a muster of *Forrestal* crewmen was being taken, fires were still burning in the ship's carpenter shop and on the hangar deck. By 2030 the fires in the 02 and 03 levels were contained, but the area was still too hot to enter. Holes were cut in the flight deck to provide access to compartments below, but only a few minutes later, *Forrestal* reported that while the fires on the 02 level were under control, firefighting was being greatly hampered by the smoke and heat. At this point medical evacuation to the *Repose* was in progress. Shortly after midnight on 30 July all the fires were finally out as the crew continued to clear smoke and cool hot steel on the 02 and 03 levels. It had been fourteen hours of hell.

Aftermath

As the adrenaline began to wear off, the bone-tired men of the *Forrestal* at last had time to ponder the events they had survived and the shipmates they had lost. They were too tired and too sick from all that they had witnessed to sleep and wandered about listlessly offering to help where they could, but when calls for volunteers went out, they raced down the passageways to man the hoses again to fight the occasional flare ups. Captain Beling said of his crew: "I am most proud of the way the crew reacted. The thing that is foremost in my mind is the concrete demonstration that I have seen of the worth of American youth. I saw many examples of heroism. I saw, and subsequently heard of, not one single example of cowardice." As the *Forrestal* steamed for Subic Bay, a memorial service for the crewmen who had given their lives for their ship and their country was held in Hangar Bay 1. During the 15-minute service of prayer and hymns, more than 2,000 *Forrestal* men listened to and prayed with Chaplains Geoffrey Gaughan and David Cooper as they paid tribute to their lost shipmates. The Marines fired three volleys in salute; they were followed by the benediction, which closed the service. After only eight days in Subic Bay, enough repairs had been made that she could make the long trip back to Norfolk for permanent repairs, where she would spend seven months. She was rebuilt from the hanger up and forward to Aircraft Elevator 4, nearly a fifth of the ship's total length and five decks. In April 1968 the *Forrestal* rejoined the fleet, but she would never to return to Vietnam.[6]

Hard Lessons

Although investigators were unable to conclusively identify the exact chain of events behind the fire, they identified potential maintenance issues with stray voltage in the LAU-10 Zuni rocket launchers. Proper procedures should have prevented this because the electrical safety pin on the Triple Ejector Rack (TER) would prevent any electrical signal from reaching the rockets, but high winds could sometimes catch the attached tags and blow them free. The backup was a "pigtail" connection of the electrical wiring to the rocket pod. They were supposed to be connected only when the aircraft was attached to the catapult ready to launch, but Navy investigators found out that this was not being done on the *Forrestal*. (Problems with faulty pigtail wiring would delay missions because the aircraft had to be removed from the launcher to correct the problem.) The inquiry found that the TER pin was likely blown free while the pigtail was connected. When the pilot transferred his systems from external to internal power, it created a power surge that caused the missile to fire. The Navy subsequently implemented safety reviews for weapons systems on board ships as well as during shipping.[7]

The Navy also instituted changes in carrier equipment. Two new firefighting chemicals were developed: Purple K powder and a "light water" system that flooded the flight deck with foam. Purple K put out fires—especially fuel-fed fires—faster than previous materials. The new light water system, called Aqueous Film Forming Foam (AFFF), used a chemically fortified agent that mixed with seawater from the existing carrier washdown systems. (Flight deck washdown systems were originally developed to create a curtain of seawater over and around the ship to protect against nuclear fallout, whereas previous firefighting foams would only work with fresh water.)

In 1968, the *Independence*, having recently completed a nine-month overhaul at the Norfolk Naval Shipyard, served as the test bed for the new systems. (She also tested using a UH-2 Seasprite helicopter to apply light water using the rotor downwash.)[8] These new systems were later installed on all carriers. Other fire safety improvements were developed, such as a heavy duty, armored forklift for use in the emergency jettisoning of aircraft (particularly heavier types such as the RA-5C Vigilante or A-3 Skywarrior) as well as heavy or damaged ordnance.

But the greatest changes occurred in the Navy's whole approach to safety at sea. The Navy still refers to the fire on the *Forrestal*, and the lessons learned, when teaching damage control and ammunition safety, and a large portion of the basic training for every sailor is dedicated to firefighting and fire prevention. Today, it is said that every Navy sailor is a firefighter first.

Turning Point and Aftermath 10

With the *Forrestal* heavily damaged and out of action, the *Constellation*, *Oriskany*, and *Intrepid* continued to hit bridges, depots, SAM sites, and airfields before the monsoon set in. On her third Vietnam deployment the *Constellation*, with a new air wing (CVW-14) embarked, accounted for four more MiG kills. On 10 August two F-4Bs from the VF-142 Ghostriders (Lieutenant [jg] Guy H. Freeborn and Ensign Robert J. Elliot; Lieutenant Commander Robert C. Davis and Lieutenant Commander Gayle O. Elie) each brought down a MiG-21 with Sidewinders. On 21 August 1967 some 80 SAMs were fired in response to major strikes from the three ships, as they hit supply depots, rail yards, and airfields. *Oriskany* also hit torpedo boat bases near Haiphong in August, sinking three vessels. The Connie scored again on 26 October when an F-4B from the VF-143 Pukin' Dogs flown by Lieutenant (jg) Robert P. Hickey Jr. and Lieutenant (jg) Jeremy G. Morris brought down a MiG-21 with a Sparrow. Her last kill of the deployment came on 30 October when a VF-142 F-4B (Lieutenant Commander Eugene P. Lund and Lieutenant [jg] James R. Borst) downed a MiG-17, also with a Sparrow. The *Coral Sea* had joined in September, but by November the monsoon rains and fog had returned so that only the A-6 Intruders flew regularly. Intensive mining efforts were made to deter supply traffic as much as possible before the upcoming Christmas bombing halt. Although MiG activity remained constant, only three MiGs fell to Navy fighters in the last three months of 1967.[1]

In December the *Ranger* arrived with the new Vought A-7 Corsair II. The A-7 bore a superficial resemblance to the F-8, but was an entirely new design. Although it was intended to replace the A-4 Skyhawk, the two aircraft continued to operate side by side during the war.

Snakeye

The Vietnam war marked the first time that the Mark 80 series of low-drag, general purpose (LDGP) bombs, which had replaced those of Korean War vintage, were used in combat.[2] Because of the growing need to be able to deliver ordnance at high speed and low altitude, retarding mechanisms were developed to quickly slow the bomb and allow the delivering aircraft to escape the blast pattern. Snakeye fin assemblies are used with the Mark 82 LDGP bombs. As soon as the bombs were dropped four retarding panels popped open, immediately creating drag and slowing the bomb down to allow the attacking aircraft to escape. The Snakeye offered greater operational flexibility because of the option for either low-drag or high-drag delivery.

Over the Christmas stand down, reconnaissance flights confirmed that a massive supply effort into the South was under way and that something big was being planned. When the New Year's truce period ended, strikes were launched to staunch the flow of men and supplies. From 2 to 11 January, aircraft from *Oriskany*, *Ranger*, and *Coral Sea* struck bridges around Hanoi and Haiphong, SAM sites, and storage depots. During that same month an incident occurred that momentarily diverted American attention from the war in Vietnam.

The *Pueblo* Incident

On 22 January 1968 the intelligence-gathering ship *Pueblo* was seized by the North Koreans in the Sea of Japan and her crew imprisoned. In response, the *Enterprise*, en route to Vietnam, was diverted to act as the flagship of Task Force 71; Task Force 71 remained in the Sea of Japan as a contingency force. By the time the *Enterprise* was relieved by the *Kitty Hawk* on 6 February, the North Vietnamese invasion of the South had begun.

Tet and Khe Sanh

On 30 January 1968, the day before the Vietnamese holiday of Tet, the Viet Cong struck at several points in South Vietnam—Nha Trang, Pleiku, Da Nang, and Qui Nhon—and even penetrated the American embassy in Saigon the next day, only to be repulsed. The Tet offensive was primarily a land campaign with little air or naval involvement. Within two weeks the communists lost thousands of men and had failed to rally the South Vietnamese people to their cause. In the northwest, however, the camp at Khe Sanh was fighting for its life. American forces had occupied Khe Sanh, only six miles from the Laotian border, since 1962. An important block to communist supplies coming south from Laos, it was reinforced by Marines in January 1967. The communists began their attacks on 21 January 1968 and for the next 71 days, in a siege reminiscent of Dienbienphu, the Marines held on. Overwhelming American air superiority, including B-52s used in a tactical role, eventually prevailed. When the siege of Khe Sanh was lifted in April, President Johnson imposed a partial bombing halt that prohibited attacks north of the 20th parallel in order to motivate the communists for projected peace talks. (Only the southernmost part of North Vietnam was below the 20th parallel.)

The Tet offensive, while a military failure, was a political victory, and served to divide American opinion on the conduct of the war. Some wanted to reinforce our troops in Vietnam, others wanted the United States to disengage. President Johnson decided that reinforcement was not the answer and felt a dialogue between the communists, the South Vietnamese, and the Americans was the only way to resolve the conflict. He announced that he would not seek reelection in 1968 and, in a misguided gesture of good faith, imposed a partial bombing halt on 31 March. With the restrictions imposed on attacking the North, tangles with MiGs were rare. (An F-8 from *Bon Homme Richard* scored against a MiG-21 on 26 June and an *Intrepid* F-8 scored again on 19 September. These were the last kills for the Crusaders in the war.) In April the *America*, an East Coast carrier, had departed for the first of three deployments to Vietnam. On 10 July an F-4J from the VF-33 Tarsiers (Lieutenant Roy Cash Jr. and Lieutenant [jg] Joseph E. Kain Jr.) brought down a MiG-21 with a Sidewinder about twenty miles northwest of Vinh, North Vietnam for *America*'s first kill of the war.[3] The year 1968 also marked the end of the venerable A-1 Skyraider's career. The propeller-driven "Spad" could no longer survive in the dangerous skies over Vietnam and was supplanted by the A-7.

The communists used the partial bombing halt to their advantage and continued to press their attacks in the South. President Johnson, in response to mounting criticism of the war at home, imposed a complete bombing halt on North Vietnam on 1 November. When Richard M. Nixon was inaugurated as president in January 1969, he was left with few options in light of the stalemate in Vietnam and the antiwar sentiment at home. During the first half of 1969, operations concentrated on South Vietnam as the bombing halt imposed the year before was observed. By February four carriers were on station: *Hancock*, *Kitty Hawk*, *Ranger*, and *Coral Sea*, but the *Hancock* left mid-month. The *Ticonderoga*, coming back on station, brought two new A-7 squadrons to Vietnam, the first equipped with the A-7B version.

The beginning of 1969 saw another major carrier flight deck fire. On 14 January the *Enterprise* was conducting exercises off Hawaii when a Zuni rocket on an F-4 ignited during startup procedures. Within minutes the fire reached major proportions, but was brought under control within three hours—28 men died and 15 aircraft were destroyed.

Withdrawal

The bombing halt continued, but on 5 June strikes into North Vietnam were launched in retaliation for the downing of an RF-8 reconnaissance Crusader. Three days later, President Nixon, during a meeting with South Vietnamese president Nguyen Van Thieu, announced that a phased withdrawal of American troops would begin. Twenty-five thousand men were withdrawn from South Vietnam by the end of August; over the next year, over 100,000 troops were pulled out. The carriers, however, remained on station, hitting targets in the northernmost region of South Vietnam, the I Corps area. In August the North Vietnamese unexpectedly released three prisoners of war and the brutal treatment suffered by the POWs was confirmed. Although air operations continued, encounters with MiGs were infrequent. In the new year, 1970, the war went on much as before as the withdrawal of the ground forces continued. The next Navy MiG kill was not until 28 March 1970, when a VF-142 F-4J (Lieutenant Jerome E. Beaulier and Lieutenant Steven J. Barkley) from the *Constellation* brought down a MiG-21 with a Sidewinder. It would be nearly two years before another MiG fell to the F-4s of the carriers.

Cambodia and Laos

In March 1970 a military coup in Cambodia deposed Prince Norodom Sihanouk while he was in Moscow seeking to reduce communist activity in his country. General Lon Nol took over the government and requested American help in dealing with Viet Cong infiltration and camps in Cambodia. By the end of April, a full-scale invasion, spearheaded by South Vietnamese troops, was under way. The campaign did not stop the communists from using Cambodia and only served to intensify antiwar protests within the United States.

Although the bombing halt of November 1968 ended attacks over North Vietnam, operations neither stopped nor diminished, they merely shifted to infiltration routes through Laos. The routes making up the Ho Chi Minh Trail were subjected to day and night surveillance using an ever increasing number of remote acoustic and seismic sensors. Occasionally, retaliatory raids were made into southern North Vietnam. By early 1971 a buildup of North Vietnamese forces in the panhandle region of Vietnam seemed to indicate that an invasion of Cambodia or Laos and South Vietnam was imminent and American help in countering this threat was needed. In February Operation Lam Son 719, a name that commemorated a Vietnamese victory over the Chinese in the 15th century, was launched as South Vietnamese troops jumped off from Quang Tri province in South Vietnam. Under American air cover, South Vietnamese units crossed into Laos. The North Vietnamese put up heavy resistance and in several areas South Vietnamese troops had to be picked up by helicopter under enemy fire. Before March ended, the last South Vietnamese troops had left Laos. Both Hanoi and Saigon claimed victory, but within weeks, the supply routes that had supposedly been disrupted were again carrying traffic south.

The anti-truck campaign in Laos ground on through 1971 as frustrated pilots from the attack squadrons faced daily risks in a futile effort to slow down the flow of supplies to the south. The fighters flew escort for photo missions and stood ready for a MiG threat that never materialized. The stage was set for major North Vietnamese efforts, and, as the new year approached, American air activity was stepped up as protective reaction raids below the 20th parallel increased. The North Vietnamese brought mobile SAMs into sites near the DMZ and during the last three weeks of 1971, ten American aircraft were lost over Laos and North Vietnam from a variety of combat measures. On 19 January 1972, during a protective reaction strike in an area where an RA-5C Vigilante and its escorts had been fired upon by antiaircraft artillery and SAMs, an F-4J from the VF-96 Fighting Falcons brought down a MiG-21 with Sidewinders, the first Navy kill since the Connie's VF-142 kill in March 1970. It was the first kill for Lieutenant Randall H. "Duke" Cunningham and Lieutenant (jg) William P. "Irish" Driscoll. Cunningham and Driscoll went on to become the only Navy aces (five or more kills) of the Vietnam war.[4]

The Easter Invasion

On 30 March 1972, the Thursday before Good Friday, three North Vietnamese divisions pushed through the DMZ to kick off the long awaited invasion of South Vietnam. Of the North's 13 regular divisions, 12 divisions, with over 120,000 troops, were eventually sent into South Vietnam. As the South's troops fell back under heavy pressure, the North used tanks in significant numbers for the first time in the war. In response, B-52 Arc Light operations expanded almost immediately and Marine air squadrons that had left South Vietnam were returned. Only the *Coral Sea* and *Hancock* were on station in the Gulf of Tonkin; the *Constellation* was on her way to Hong Kong, but was recalled. The *Kitty Hawk* soon joined, while the *Enterprise* was on duty in the Indian Ocean. The *Saratoga*, the first ship to be redesignated under the "CV concept," was recalled from the Atlantic Fleet in April and others were transiting the Pacific. (This was the beginning of a return to general purpose aircraft carriers. See chapter 7.)

Linebacker

For two years, in public and secret meetings with the North Vietnamese, Dr. Henry Kissinger, President Nixon's National Security Advisor, had tried to get the North to agree to a cease-fire. Le Doc Tho, the North's representative, played for time, alternately agreeing, then disagreeing with American negotiators. In May, after the North had unleashed a major attack, Kissinger, after a frustrating session with Le Doc Tho, reported to the president that the communists were intransigent. Faced with an upcoming summit meeting with Soviet leader Leonid Brezhnev, presidential elections, and the strong possibility of public outrage at home, President Nixon decided to take the gamble and ordered the mining of North Vietnamese ports to cut off the supply of material into the South. On 9 May A-6s and A-7s from the *Coral Sea* mined the waters of Haiphong Harbor. Later, on 11 May, aircraft from *Coral Sea*, *Midway*, *Kitty Hawk*, and *Constellation* laid additional mine fields in the ports of Thanh Hoa, Dong Hoi, Vinh, Hon Gai, Quang Khe, and Cam Pha, as well as the Haiphong approaches.[5]

On 6 May one of the most active dogfights of the war occurred when two MiG-21s were downed by *Kitty Hawk*'s VF-114 (Lieutenant Robert G. Hughes and Lieutenant [jg] Adolph J. Cruz; Lieutenant Commander Kenneth W. Pettigrew and Lieutenant [jg] Michael J. McCabe) and two MiG-17s from the *Coral Sea*. On 8 May VF-96's Lieutenant Cunningham and Lieutenant (jg) Driscoll scored their second kill, a MiG-21, but 10 May brought the most intense aerial combat of the war, when Navy flyers brought down eight MiGs. Over Haiphong, Cunningham and Driscoll shot down three MiG-17s for the first triple downing of enemy MiGs by one plane during the war. These three MiGs, with their previous kills, made them the first aces of the Vietnam War. Three other kills were scored by planes of VF-96. Two MiG-17s were downed by Lieutenant Matthew J. Connelly III and Lieutenant Thomas J. J. Blonski, while Lieutenant Steven C. Shoemaker and Lieutenant (jg) Keith V. Crenshaw downed another MiG-17. Lieutenant Curt Dose and Lieutenant Commander James McDevitt from *Constellation*'s other F-4 squadron, the VF-92 Silverkings, shot down a MiG-21 while another MiG-17 was shot down by VF-51 off *Coral Sea*.

By July six carriers were off Vietnam: *America*, *Hancock*, *Kitty Hawk*, *Midway*, *Oriskany*, and *Saratoga*. This was the greatest number of carriers on station during the war. The carriers hit targets in the North and South, leaving the *Hancock* to concentrate on truck and troop positions in the South, while A-7s and A-6s from the "big deck" carriers hit AAA sites, bunkers, and supply depots. *Saratoga*, an Atlantic Fleet carrier, had left Mayport in April for her only deployment to Vietnam. On 21 June, the last day of her first line period, two of her F-4Js from the VF-31 Tomcatters attacked three MiG-21s over North Vietnam. Dodging four SAMs, the Phantom, flown by Commander Samuel C. Flynn Jr. and Lieutenant William H. John, shot down one of the MiG aircraft with a Sidewinder.[6] On 10 August 10 *Saratoga* scored again when an F-4J flown by Lieutenant Commander Robert E. Tucker Jr. and Lieutenant (jg) Stanley B. Edens from her other fighter squadron, the VF-103 Jolly Rogers, shot down a MiG-21 with a Sparrow missile during a night interception, marking the first and only night MiG kill by the Navy.[7]

During the five and one-half months of Linebacker I, the Navy flew well over half of the total sorties over North Vietnam, most these were in the "panhandle" area between Hanoi and the DMZ. Tactical air operations were most intense between July and September, with most sorties devoted to either armed reconnaissance missions or strikes. The armed reconnaissance missions were usually against targets of opportunity near Hanoi, Haiphong, and the Chinese border. Strike operations were preplanned, usually against fixed targets. On 18 May, when the Uong Bi electric power plant near Haiphong was struck, it marked the beginning of strikes against targets—such as power plants, shipyards, and the Haiphong cement plant—that were previously avoided.

In June 1972 a Marine fighter squadron, the VMFA-333 Fighting Shamrocks, had deployed on the *America* for her third and last deployment to Vietnam. It was during this deployment that VMFA-333 got the only all-Marine air-to-air MiG kill of the war when Major Lee T. Lasseter and Captain John D. Cummings shot down a MiG-21 near Hanoi. Both aircraft in the flight were damaged by flak and an SA-2 SAM. Close to running out of fuel and with one of the pair on fire from the SAM hit, the crews ejected just south of Haiphong Harbor.[8]

The principle Navy attack aircraft involved were the A-7s and A-6s. (The A-7s accounted for roughly 60 percent of Navy attack sorties while the A-6s concentrated on night attacks, which were about a fourth of the overall Navy effort.)[9] The strikes against the North had been given the name Freedom Train, but now, with the mining

Artist concept of the *United States* (CVB-58), the basis for the *Forrestal* design.

Forrestal under construction at Newport News Shipbuilding, April 1954. (Newport News Shipbuilding)

A view of the *Forrestal*'s port after 5" guns at Newport News Shipbuilding, September 1955. (Newport News Shipbuilding)

The *Forrestal* backing away from her outfitting pier, August 1954. (Newport News Shipbuilding)

Top: Another view of the *Forrestal* leaving her outfitting pier. The blimp overhead is flying channel guard as she makes her way out of Hampton Roads. (Newport News Shipbuilding)

Center: *Forrestal* during her sea trials, January 1956. (Newport News Shipbuilding)

Right: A view of *Forrestal*'s island at Newport News Shipbuilding, September 1955. (Newport News Shipbuilding)

A starboard side view of the *Forrestal* in 1955.

During *Forrestal*'s shakedown in March 1956, an FJ-3 Fury launches from the amidships catapult while another from VF-21 and an F2H-3 Banshee are being readied for launch from the bow catapults.

An F7U-3 Cutlass from VA-83 prepares to launch from the *Forrestal* while another moves into position on the starboard catapult, March 1956.

Forrestal during sea trials, January 1956. (Newport News Shipbuilding)

An A3D Skywarrior prepares to launch from the *Forrestal* during the 1960s.

Right: Two AD-5W Skyraiders from VAW-12 fly over the *Forrestal* during her deployment with the Sixth Fleet in the Mediterranean, April 1960.

Center: Crewmen on deck on board the *Intrepid*, *Saratoga*, and *Independence* spell out a message commemorating the 50th anniversary of naval aviation, 1961.

Bottom: *Saratoga* at anchor in Hampton Roads during the International Naval Review, June 1957.

The *Ranger* during her sea trials, July 1957. (Newport News Shipbuilding)

Four F8U Crusaders launch in quick succession from the *Ranger* as she operates in the Pacific, 1958.

The *Ranger* under way in the Pacific, May 1975. She retains two of her 5" guns. The circular markings on her flight deck are helicopter landing spots.

A3D Skywarriors, A4D Skyhawks, F3H Demons, and F8U Crusaders are seen on the flight deck of the *Independence* during her shakedown cruise in April 1959. Note the interim pattern on the landing path.

The *Kitty Hawk* under way in 1961.

The *Constellation* under construction at the Brooklyn Navy Yard, 1960.

A Terrier missile launches from the *Constellation*, 1962.

The combat stores ship *Niagara Falls* providing underway replenishment for the *Constellation* in the South China Sea, December 1979. The guided-missile cruiser *Leahy* is on her starboard side.

The *America* departing Newport News Shipbuilding after delivery, November 1964. (Newport News Shipbuilding)

The *America* under way, 1983.

Right: An RA-5C Vigilante prior to launch from the *America*, 1966–67.

Center: An F-4 Phantom recovers on board the *John F. Kennedy* in the 1970s.

Bottom: Sailors man the rails of the *Independence* as she is welcomed by a fire tug while entering New York Harbor. She was en route to a 4th of July visit to the 1964 World's Fair.

Top: An F-14A Tomcat is ready for the first catapult test launch from the *Forrestal*, June 1972. New Mark 7 Jet Blast Deflectors had been recently installed to handle the heat blast from the Tomcat's engines.

Right: Steam rises from the catapults on the flight deck of the *John F. Kennedy* as an A-4 Skyhawk is readied for launch, 1968–69.

Bottom: A flight deck crew member checks the bridle on an F-4 on board the *John F. Kennedy*, late 1968. Bridles were the standard method of hooking up to the catapult until replaced by nose-tow bars.

An S-3 Viking from VS-30 lands on the *John F. Kennedy* during Summer Pulse 2004, an exercise intended to demonstrate the Navy's ability to deploy seven carriers simultaneously.

A starboard side view of the *America* during her deployment to Vietnam from April to December 1968. Note the distinctive shape of her funnel.

Top: Various aircraft, including A-6 Intruders, F-14 Tomcats, A-7 Corsairs, SH-3 Sea King helicopters, E-2C Hawkeyes, and EA-6B Prowlers, are spotted on the flight deck of the *Kitty Hawk* in January 1980.

Center: A U-2 reconnaissance aircraft on the flight deck of the *America*, October 1984.

Left: A C-130 Hercules on the deck of the *Forrestal* during carrier suitability trials, October 1963.

Top: VA-196 A-6A Intruders from *Constellation* dropping Mark 82 bombs over Vietnam during her May 1968 to January 1969 deployment to Vietnam.

Left: An A-7 Corsair from *Kitty Hawk*'s VA-195 bombs the Hai Duong railway and highway bridge in North Vietnam during Operation Linebacker I, May 1972.

Bottom: An A-7 Corsair from VA-147 is launched from one of *Ranger*'s waist catapults, during operations in the Tonkin Gulf, January 1968.

operations, the air offensive was given the code name Linebacker. The effort was more intense than previous Rolling Thunder operations and the communists were pushed back by the South Vietnamese backed by American air power. The North began to show more signs of willingness to negotiate.

Racial Unrest

The Vietnam War took place during a time of great social turmoil in American society as the nation struggled with the issues of race relations and civil rights. Against this backdrop was the divisive issue of the war itself. Because of the draft, many African Americans felt the war was being fought disproportionally by minorities and the economically disadvantaged and many had enlisted in other services to avoid ground combat in Vietnam. Racial tension was endemic within all of the services as the war dragged on with no clear resolution in sight. For the ships operating off Vietnam, extended deployments and long line periods had produced considerable strain on their crews. On board the *Kitty Hawk* her commanding officer, Captain Marland W. Townsend, allowed black sailors to stay in their own berthing spaces without other races, which had the unintended consequence of segregating black sailors from their shipmates. Shortly after 1830 on 11 October 1972, a series of incidents on the mess decks led to fighting between blacks and whites that spread to other areas of the ship, including sick bay and the flight deck. The ship's marine detachment, following regulation crowd control procedures, conducted patrols to restore order and attempted to prevent groups of more than three sailors from congregating, while maintaining a 12-man reaction force. Some black sailors interpreted these efforts as racist and armed themselves with aircraft tie-down chains. Captain Townsend and the XO, Commander Benjamin W. Cloud, who was black, addressed the rioters several times. Eventually tempers cooled and order was restored. There were 47 reported injuries, but the Hawk continued to support Linebacker operations, resuming strikes on the morning of 13 October.[10]

In November another racial incident occurred on board the *Constellation* as she was outfitting for her next deployment to Vietnam. Her crew worked long hours in the hot ship, which was not air conditioned, and constant construction noise made it impossible for many crewmen to sleep. As tension mounted, some black sailors launched a sit-down strike to protest their conditions. The Connie's captain ordered the dissidents off his ship, where they were joined ashore by other angry black sailors, who were convinced that their mistreatment was racially motivated. With all the media attention that followed, they became symbols of the Navy's racial problems. The CNO, Admiral Elmo Zumwalt, and Navy Secretary John Warner personally intervened in the negotiations that ended the sit-down strike. Congressional inquiries that followed concentrated on the role of a few agitators and blamed growing overall permissiveness and lax discipline within the Navy as the Vietnam War wound down. Admiral Zumwalt, for his part, instituted Navy-wide training programs to encourage tolerance, racial awareness, and harmony. Many of these programs were subsequently eliminated by his successor, Admiral James Holloway, in favor of minority recruitment programs, greater emphasis on education (including remedial education), and more promotion opportunities.[11]

Linebacker II and the Paris Peace Talks

The October 1972 Paris peace talks produced little; the North Vietnamese were again unwilling to agree to a cease-fire once President Nixon had halted bombing north of the 20th parallel. Instead, they used the respite to resupply and rebuild their defenses. In December their delegation walked out of the negotiations. With few choices left, President Nixon ordered a maximum bombing effort against the Hanoi-Haiphong area. Linebacker II began on 18 December with waves of B-52s and F-111s. Before the B-52s struck, the Navy sent in airfield suppression strikes, particularly at Kep, the main MiG base in the Hanoi area. By the end of December, the North Vietnamese showed a willingness to negotiate in earnest and returned to the peace talks. On 23 January 1973 a cease-fire was announced. America would have "peace with honor"—for the time being.

Eagle Pull and Frequent Wind

The cease-fire did not end the war itself. North Vietnam and the United States had agreed on ending the fighting, returning prisoners of war, and clearing mines from North Vietnamese harbors, but Cambodia and Laos continued fighting against communist insurgents. While Americans continued withdrawal from Vietnam and the POWs returned home, strikes were flown into Laos. On 28 January the *Ranger* and *Enterprise* flew 81 sorties against line-of-communication targets in Laos, overflying South Vietnam between Hue and Da Nang. On 11 February the *Constellation* and *Oriskany*, operating from the "new" Yankee Station (it had been moved south to a point off the northern coast of South Vietnam) flew strikes into southern Laos. On 25 February the *Ranger* and *Oriskany* flew combat support missions over Cambodia.[12] The capital of Cambodia, Phnom Penh, was under siege; its only access routes to the outside were through convoys of ships along the Mekong River. The communists protested American strikes supporting the convoys as violations of the Paris peace accords; the American response was to halt mine-clearing operations, which had begun in February under Operation End Sweep. By June all sides agreed to better enforcement of the cease-fire and by August American strikes had stopped. Laos accepted a coalition government and Cambodia was left to fend for itself.[13] The communists had only to wait until American support was too little to make a difference. The end came in 1975. Cambodia fell first, as the Khmer Rouge, along with thousands of North Vietnamese and Viet Cong troops, put the final strangle hold around Phnom Penh. On 12 April Marine helicopters operating off the *Hancock* evacuated the American embassy in Operation Eagle Pull. The communists swept up city after city in South Vietnam and by the end of the month were at the outskirts of Saigon. During Operation Frequent Wind, thousands of Americans and their South Vietnamese allies were evacuated during the "night of the helicopters," 29 to 30 April. North Vietnamese tanks rolled through the gates of the Presidential Palace in Saigon the next morning.[14]

The *Mayaguez* Incident

After the fall of South Vietnam, there was one final act in America's involvement in Southeast Asia. The American merchant ship SS *Mayaguez* was steaming in international waters en route to Thailand when she was boarded on 12 May 1975 and her crew seized by the new communist government of Cambodia. Following so closely on the humiliation of the fall of Saigon, President Gerald Ford, who had taken over from President Nixon less than a year before, felt he had to move quickly. On 15 May a force of Marines was landed by helicopter on Koh Tang, a small island off Cambodia where the crew was believed to be held. Carrier aircraft strikes were sent in to support the Marines. Two of the helicopters were shot down by the Cambodians before it was learned that the crew had been released. Casualties included 15 men killed, 3 missing, and 50 wounded. It was a bitter footnote to a long and costly war.[15]

11 Evolution

Just as the first generation of super carriers had been adapted to accommodate new technology and developing operational concepts, so too were new aircraft and weapons introduced to meet emerging requirements. Some of these new aircraft would, in turn, require changes to be made to the carriers themselves.

F-14 Tomcat

In the early 1960s both the Air Force and Navy were looking for new fighter designs. The Air Force Tactical Air Command wanted a fighter-bomber for deep strike and interdiction missions to replace the F-105 Thunderchief, while the Navy wanted a high-altitude, long-range interceptor to defend its carriers from anti-ship missiles launched from Soviet jet bombers and submarines. Despite the difference in missions, Defense Secretary McNamara directed the services to adopt a common aircraft for what became the Tactical Fighter Experimental (TFX) program. After numerous proposals, Secretary McNamara selected General Dynamics, primarily because of the higher degree of commonality between the Air Force and Navy versions. The F-111A for the Air Force and the Navy F-111B shared the same airframe structural components with variable-geometry wings, side by side crew seating in an escape capsule, afterburning TF30 turbofan engines, and the long-range AWG-9 radar and AIM-54 Phoenix missile weapons system. Because it had no experience with carrier aircraft, General Dynamics partnered with Grumman to develop the Navy version. The F-111B nose was shorter to fit on carrier elevators, and had longer wingtips to improve on-station endurance. Both designs proved to be overweight and underpowered and only seven F-111Bs were built.[1] Meanwhile experience with air-to-air combat over North Vietnam had shown the need for greater maneuverability and the F-111B had turned out to be less maneuverable than the F-4 Phantom it was intended to replace. Grumman had been studying various improvements and alternatives to the F-111B and this would ultimately result in the F-14 Tomcat. Grumman had a tradition of naming fighters after various cats, the Wildcat, Hellcat, Panther, Cougar, Tiger, etc., and the name Tomcat was selected to honor Vice Admiral Thomas Connolly, Deputy Chief of Naval Operations for Air Warfare. The name also honored Admiral Thomas H. Moorer, who was the Chief of Naval Operations at the time.[2]

Grumman's design used the same TF30 engines from the F-111B, though the Navy planned on replacing them with improved engines later. The large AWG-9 radar and AIM-54 Phoenix missiles were also retained although the Tomcat incorporated an internal M61 Vulcan 20mm cannon, as well as providing for Sidewinder and Sparrow air-to-air missiles and bombs. Ironically, even though lighter than the F-111B, the Tomcat was still the largest and heaviest fighter to operate from a carrier. Thanks to its variable geometry wings, however, the F-14 was more maneuverable than the smaller F-4. Having first flown in December 1970, the Tomcat made its first deployment in 1974 on board the *Enterprise*. The F-14 was the Navy's primary air superiority fighter, fleet defense interceptor, and tactical reconnaissance platform for many years. In service the Tomcat was known as the "Turkey" for its appearance when landing. The F-14 had a slower approach speed than the F-4 and corrections required greater control movement on the part of the pilot. With the horizontal stabilators moving differentially, the spoilers rising on the wing, and its flapping and rocking, the resemblance to a turkey was only natural. (The Grumman TBF Avenger of World War II was also known as the "Turkey," probably for similar reasons.)

The Hughes AIM-54 Phoenix was a radar-guided, long-range air-to-air missile that was the only U.S. long-range air-to-air missile. (Before the introduction of the Phoenix, most other American fighter aircraft relied on the smaller, less-expensive, AIM-7 Sparrow medium-range missile.) Combined with the AWG-9 radar, which was capable of tracking up to 24 targets simultaneously in Track-While-Scan mode, up to six priority targets could be engaged by Phoenix missiles at one time. The F-14 was the only aircraft to carry the Phoenix and, although technically very capable, the great range of the Phoenix caused it to be rarely used in combat because of the rules of engagement, which would require positive visual identification of targets.[3]

As the largest and heaviest U.S. fighter to fly from an aircraft carrier when it entered service, changes had to be made to the carriers to accommodate it. The Jet Blast Deflectors, for instance, had to be enlarged and liquid cooled to be able to handle the heat generated by the TF30 engines when they went into afterburner prior to launch.[4]

The Tomcat was continually improved over its service life. In 1987 some F-14As were upgraded with General Electric F110-400 engines. Initially designated F-14A+, they were redesignated as F-14Bs in 1991. The F-14D variant was developed at the same time and used the same F110-400 engines with newer digital avionics systems such as the Digital Flight Control System (DFCS) that greatly improved the F-14's handling when flying at high angles of attack or in air combat maneuvering. With the retirement of the RA-5C Vigilante and RF-8G Crusader reconnaissance aircraft, the Tactical Airborne Reconnaissance Pod System (TARPS) was developed in the late 1970s for the F-14. Some F-14As and all F-14Ds were modified to carry the pod, which was primarily controlled by the radar intercept officer (RIO). Although there was some concern that fighter crews could not replace dedicated photo pilots, in practice this did not prove to be a problem. (The TARPS was upgraded with a digital camera in 1996 [the TARPS-DI] and further updated starting in 1998 with the TARPS Completely Digital [TARPS-CD] to provide real-time transmission of imagery.)

During Operation Desert Storm, most air-to-ground missions were left to A-7 and F/A-18 squadrons, while the F-14 focused on air defense operations, but in the 1990s, the Low Altitude Navigation and Targeting Infrared for Night (LANTIRN) pod system was added to enable precision ground-attack missions. The last Tomcat left the fleet in September 2006, its role taken over by the F/A-18E/F Super Hornet.[5]

F/A-18 Hornet and Super Hornet

The Navy had wanted a multirole aircraft to replace the A-4 Skyhawk, the A-7 Corsair, and the remaining F-4 Phantoms, as well to complement the F-14 Tomcat. Congress had mandated that the Navy find a lower-cost alternative to the F-14, but both Grumman and McDonnell Douglas had submitted proposals that were nearly as expensive as the F-14. In 1973 Defense Secretary James Schlesinger ordered the Navy to evaluate the competitors in the Air Force Lightweight Fighter (LWF) program, the General Dynamics YF-16 and Northrop YF-17. The Air Force adopted the F-16, but the Navy preferred the twin-engine F-17. Northrop partnered with McDonnell Douglas to develop an essentially beefed up version of the F-17 that became the F-18 Hornet, which first flew in 1978. Fighter and attack versions of the basic aircraft subsequently emerged as the F/A-18A single-seat version, which entered service in 1983, and the F/A-18B two-seat version, which was primarily used for training. The F/A-18 first saw combat action in April 1986 against Libyan air defenses during Operation Prairie Fire and an attack on Benghazi as part of Operation El Dorado Canyon. In 1987 the Hornet was upgraded to the F/A-18C (single-seat) and F/A-18D (dual-seat) versions. (The Hornet is currently in service with the Navy and Marines as well as Australia, Canada, Finland, Kuwait, Malaysia, Spain, and Switzerland.) When the Navy A-12 Avenger II program, intended to replace the A-6 Intruder, was canceled, the Super Hornet was selected to fill the gap. Designed and initially produced by McDonnell Douglas, the Super Hornet first flew in 1995. Full production began in 1997, after the merger of McDonnell Douglas and Boeing. The F/A-18E single-seat and F/A-18F dual-seat variants are essentially larger and more advanced derivatives of the F/A-18C and D Hornet. The Super Hornet can carry up to five external fuel tanks and can be configured as an airborne tanker to replace the KA-6D. Entering service in 1999, the Super Hornet began replacing the F-14 Tomcat in 2006, and serves alongside the original Hornet in carrier air wings.

S-3 Viking

The Lockheed S-3 Viking, a four-seat, twin-engine jet, entered service in 1974 as a replacement for the S-2 Tracker to provide antisubmarine warfare capability for the big deck carriers under the "CV" concept. Because of the low-pitched sound of its General Electric TF-34 turbofan engines, it was nicknamed the "Hoover" after the vacuum cleaner brand. The first operational cruise of the S-3A was in 1975 on board the *John F. Kennedy* with the antisubmarine squadron VS-21, the Fighting Redtails. The upgraded S-3B version followed in 1987. With the collapse of the Soviet Union and the breakup of the Warsaw Pact, the submarine threat was regarded as much reduced, and, much like the earlier A-3, the S-3's mission focus started shifting in the late 1990s to other roles, such as surface warfare and aerial refueling, as well as providing electronic warfare and surface surveillance capabilities to the carrier battle group. (The VS squadrons were also to become "sea control" squadrons in October 1993, although the VS designation was the same.) The ES-3A Shadow was a carrier-based electronic intelligence (ELINT) aircraft that replaced the EA-3B starting in 1991. The US-3A was a Viking modified as a Carrier Onboard Delivery (COD) aircraft, with a capacity for six passengers or 4,680 pounds of cargo; it was retired in 1998. The S-3B saw extensive service in the 1991 Gulf War, in attack, tanker, and ELINT roles, and launching ADM-141 Tactical Air Launched Decoy (TALD) missiles, as well as participating in the Bosnian operations of the 1990s and in Operation Enduring Freedom in 2001. The S-3 was retired from carrier service in 2009 when VS-22, the last S-3B squadron, was decommissioned.[6]

SH-60 Seahawk

The Sikorsky SH-60 Seahawk, based on the Army's UH-60 Black Hawk twin-turboshaft engine, multi-mission helicopter, entered service in 1984 to replace the Kaman SH-2 Seasprite as the new Light Airborne Multi-Purpose System (LAMPS) platform. The SH-60B maintained 83 percent commonality with the UH-60A but with changes to adapt it to shipboard service. These included corrosion protection, more powerful T700 engines, single-stage oleo main landing gear, fuselage structure replacing the left side door, two weapon pylons, and moving the tail landing gear 13 feet forward to reduce the footprint for shipboard landing. Other changes included larger fuel cells, an electric blade folding system, folding horizontal stabilators for storage, and a 25-tube sonobuoy launcher on the left side, as well as an emergency flotation system in the stub wing fairings of the main landing gear. Following the SH-60B, the Navy developed the SH-60F to replace the SH-3 Sea King as the primary antisubmarine warfare (ASW) aircraft for carrier battle groups. The SH-60F, which first flew in 1987, is unofficially known as the "Oceanhawk." The SH-60F can carry Mark 46, Mark 50, or Mark 54 torpedoes and has the option of carrying fuselage-mounted machine guns (the M60D, M240D, and GAU-16) for self-defense. Other versions include the HH-60H, the primary combat search and rescue (CSAR), naval special warfare (NSW) and anti-surface warfare (ASUW) helicopter, and the MH-60S "Knighthawk" on board amphibious assault ships and fast combat supply ships for troop transport, vertical replenishment (VERTREP) and search and rescue (SAR). The MH-60R is designed to combine the features of the SH-60B and SH-60F.[7]

Maverick

Development of the Maverick tactical air-to-ground missile began in the 1960s as the Air Force was looking for a replacement for the Bullpup. The Maverick has a modular design that allows different combinations of guidance packages and warheads (either shaped charge or penetrating blast fragmentation) to be attached to the rocket motor section. With its delta wings and cylindrical body, it is reminiscent of the Phoenix in appearance. The Maverick has been continually improved over the years and can be launched from a variety of Navy tactical aircraft, including the Skyhawk, Phantom, Intruder, Corsair, Hornet, and Super Hornet.[8]

Paveway

The Paveway series of guided bombs was developed by the Air Force in the 1960s. (Pave is sometimes used as an acronym for precision avionics vectoring equipment. The term has often been combined with other words to describe specific guidance systems. Paveway has become a general term for laser-guided bombs [LGB].) Paveway kits attach to a variety of bombs with laser seekers, guidance and control electronics, and front control canards and rear wings for guidance and stability. The weapon homes in on laser energy—the "sparkle"—reflected off the target from a laser designator. In the 1970s the original Paveway series gave way to the Paveway II series. A new generation of weapons, Paveway III, entered service in the 1980s. Navy versions (based on the Mark 80 series low drag bombs) that could be carried by the F-14 and F/A-18 included the Paveway II GBU-10 (2,000-pound Mark 84), GBU-12 (500-pound Mark 82), GBU-16 (1,000-pound Mark 83), and the Paveway III GBU-24 (2,000-pound Mark 84).[9]

Harpoon

The Harpoon, an all-weather, over-the-horizon, anti-ship cruise missile, entered service in 1977 and has ship, submarine and air-launched versions. (Since the system was originally intended to attack surfaced submarines, the name Harpoon was chosen because the missile was designed to kill "whales," a Navy slang term for submarines.) It uses mid-course guidance with a radar seeker to attack surface ships and its low-level, sea-skimming cruise trajectory and 1,500-pound warhead ensure high survivability and effectiveness. The AGM-84 air-launched version could be carried by patrol aircraft, such as the P-3 Orion and S-3 Viking, as well as the A-6 Intruder and F/A-18. In the 1980s the improved Stand-off Land Attack Missile (SLAM) version was developed to provide an all-weather capability against high value land targets.[10]

HARM

The AGM-88 High-speed Anti-Radiation Missile (HARM) is a tactical, air-to-surface missile originally developed as a replacement for both the Shrike and Standard ARM anti-radiation missiles. The HARM missile deployed in late 1985 with VA-72 and VA-46 on the *America* and was soon used against a Libyan SA-5 site in March 1986 and then in Operation Eldorado Canyon in April. The HARM was used extensively for Desert Storm in 1991. (The brevity code "Magnum" is called out over the radio when a HARM is launched. During the Gulf War, if an aircraft was lit up by an enemy radar, a bogus Magnum call was often enough to convince the enemy radar operators to shut down. This trick was also used in Serbia in 1999.)[11]

Skipper II

The AGM-123 Skipper II, which entered service in 1985, is a short-range missile for precision strikes designed by China Lake as a cheap anti-ship missile. It is made up from a Mark 83 1,000-pound bomb, fitted with a Paveway laser guidance kit, and an attached Mark 78 rocket motor that fires when the weapon is dropped. The greater range compared to a free-fall bomb gives the delivering aircraft greater standoff from SAMs and AAA. It could be carried by the A-6E Intruder, A-7 Corsair, and the F/A-18. (There is no Skipper I, the II refers to the Paveway guidance.)[12]

JDAM

The Joint Direct Attack Munition (JDAM) is a guidance kit that converts existing unguided "dumb" bombs into precision-guided "smart" weapons. The tail section contains an Inertial Navigation System (INS) and a Global Positioning System (GPS). JDAM improves the accuracy of unguided bombs in any weather conditions and could be dropped from Navy fighter-attack aircraft such as the F/A-18 and F-14. JDAM comes in three sizes: the 2,000-pound GBU-31 using the Mark 84, the 1,000-pound GBU-32 using the Mark 83, and the 500-pound GBU-38 using the Mark 82.[13]

AMRAAM

The AIM-120 Advanced Medium-Range Air-to-Air Missile, or AMRAAM (pronounced "am-ram") was designed with the same form and fit factors as the Sparrow. Unlike the semi-active radar Sparrow, it is a modern fire-and-forget active radar system that has replaced both the Sparrow and the Phoenix. The first combat use of the AMRAAM was in December 1992 when an Air Force F-16 shot down an Iraqi MiG while enforcing the no-fly-zone. The continued success of the missile has led to its nickname of "slammer." The AIM-120 is used on the F/A-18. (When an AMRAAM is launched, pilots use the brevity code "Fox Three." The code "Fox One" is for a semi-active radar-guided missile [e.g., the Sparrow], "Fox Two" for an infrared-guided missile [e.g., the Sidewinder], and "Fox Three" is for an active radar-guided missile.)[14]

JSOW

The AGM-154 Joint Standoff Weapon (JSOW) is a joint Navy and Air Force medium-range precision-guided weapon, especially for use against defended targets from outside the range of anti-aircraft defenses. It is a launch and leave weapon with a Global Positioning System (GPS)/Inertial Navigation System (INS) and can be used day or night and in bad weather. Different versions use different terminal guidance, either GPS or infrared, and deliver different submunitions designed to attack a variety of targets. The AGM-154C (JSOW C) uses an infrared seeker for terminal guidance. It was first used in combat in December 1998 and has been used in Operations Desert Fox, Southern Watch, Enduring Freedom, and Iraqi Freedom.[15]

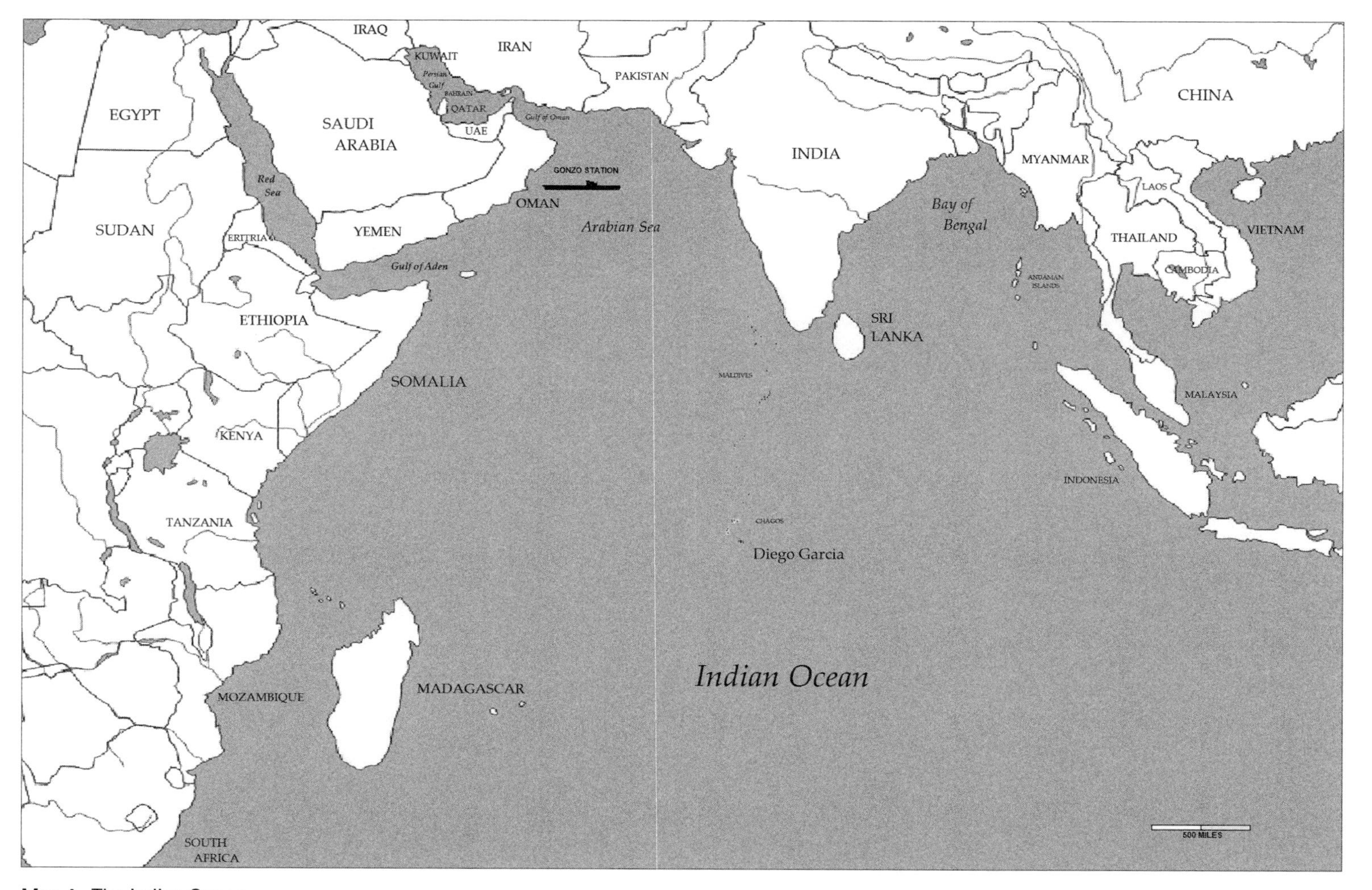

Map 4. The Indian Ocean

Map adapted by the author from http://d-maps.com/carte.php?&num_car=4318&lang=en.

12 Challenges

As American involvement in Vietnam was drawing to a close, the Navy, despite declining resources, continued to meet commitments in other parts of the world. In the troubled Middle East, carriers were often called upon to provide a stabilizing presence and be ready to intervene if need be, as happened in Cyprus in July 1974 when fighting broke out between Turkish and Greek Cypriot factions. The American ambassador to Cyprus, Roger Davies, requested the evacuation of U.S. citizens, which was carried out by a Marine helicopter squadron operating from the *Inchon*, while the *Forrestal* provided air cover.[1] In the decades that followed, American concern over its critical dependence on foreign sources of oil would require an ongoing commitment of naval forces to protect our national interests. As always, even when not engaged in actual combat, carrier operations were demanding and dangerous.

The *Belknap* Incident

On the night of 22 November 1975 the *John F. Kennedy* task group was operating in the Mediterranean east of Sicily. As the guided-missile cruiser *Belknap* maneuvered to take her station during the carrier's last recovery, she collided with the *Kennedy*'s port side. Within ten minutes firefighters had contained the fuel fire that flared up on the *Kennedy*, but a space below continued to burn for several hours and heavy smoke forced the evacuation of the carrier's fire rooms. As she went dead in the water, all flights (except the SH-3s from HS-11 supporting rescue and relief efforts) were diverted to the Naval Air Facility at Sigonella. On the *Belknap*, the situation was much worse; the carrier's overhanging angled deck ripped into *Belknap*'s superstructure from her bridge aft as the cruiser passed beneath it. JP-5 fuel from ruptured lines in the carrier's catwalk sprayed onto severed electrical wiring on what was left of the *Belknap*'s superstructure. Within minutes her entire amidships was an inferno. Ammunition from the ready storage locker began to cook off, endangering the destroyers *Bordelon* and *Claude V. Ricketts* that had come alongside to assist the stricken cruiser. Meanwhile the frigate *Pharris* moved in on the carrier's port side to assist in the firefighting. Seven men were killed on board the *Belknap* and one died of smoke inhalation on the carrier.[2] The next day the *Kennedy* was deemed operationally capable and resumed flight operations using a manually operated visual landing aid system (MOVLAS) rigged on the starboard side abreast the island on 24 November.

The Iran Hostage Crisis

In October 1977 demonstrations against Shah Mohammad Reza Pahlavi of Iran, the pro-Western leader who would become the last Persian monarch, developed into civil resistance. This resistance, which was partly secular and partly religious, intensified the following January and between August and December of 1978 the country was paralyzed by strikes and demonstrations, leading to the Shah's exile in January 1979. Weeks later, the religious leader Ayatollah Khomeini returned to Tehran to be greeted by several million Iranians. The regime collapsed in February when guerrillas and rebel troops overwhelmed troops loyal to the Shah in armed street fighting. In April, by a national referendum, Iran voted to become an Islamic Republic under a democratic-theocratic constitution that made Khomeini Supreme Leader the following December. Meanwhile, in October 1979, President Jimmy Carter had allowed the Shah, who was ill with cancer, into the United States for treatment. On 4 November 1979 a group of Islamist students and militants took over the American embassy in Tehran in support of the Iranian Revolution. The tense situation deteriorated when the U.S. attempted to free the 52 American hostages on 24 April 1980. The rescue attempt, Operation Eagle Claw, proved to be a disaster. The complex plan called for a minimum of six helicopters to arrive at an initial rendezvous known as Desert One, and from there proceed on to the objective. Even though eight helicopters were sent, two could not navigate through a sand storm and aborted. Of the six that arrived at Desert One, one had a damaged hydraulic system. At the request of the commanders on the scene, the mission was aborted, but as the aircraft prepared to leave Iran, one of the helicopters collided with a C-130 loaded with fuel, resulting in the death of eight Americans. American prestige suffered greatly and the debacle at Desert One marked both a turning point for relations between Iran and the United States and the end of President Carter's hopes for a second term. The crisis ended with the signing of the Algiers Accords in Algeria on 19 January 1981, and the hostages were released into U.S. custody the following day, only minutes after Ronald Reagan was sworn into office as president.[3]

Gonzo Station

During the Iran hostage situation and later developments in the Persian Gulf, the Navy maintained a continuous carrier presence to be ready to respond quickly to any developments. Gonzo Station was the term used to designate an area of carrier operations in the Indian Ocean during the Iranian hostage crisis and later the so-called tanker war between the U.S. and the new Islamic Republic of Iran. (Gonzo Station was actually an acronym for Gulf of Oman Naval Zone of Operations, not to be confused with the term "gonzo journalism" that is often used to describe the subjective first person stream-of-consciousness style of writing associated with Hunter S. Thompson.)[4] Several carriers served on Gonzo Station, including the *Ranger*, *Midway*, *Enterprise*, *America*, *Independence*, *Constellation*, *Coral Sea*, *Nimitz*, and *Eisenhower*. Replenishment ships normally rotated on and off-line in order to resupply the carriers in what became the largest American naval force dispatched to the Indian Ocean since World War II. While the carriers on Gonzo Station normally rotated on three-month cruises before being relieved by another task force, the lack of suitable ports in the region meant that there were extremely long periods at sea. (Gonzo Station was also important because of Soviet interest in Somalia, strategically located on the Horn of Africa near the mouth of the Red Sea. From the 1970s until he was deposed in the early 1990s by a coalition of armed opposition groups, the Marxist-leaning dictator of Somalia, Mohamed Siad Barre, first courted the Soviet Union, but switched allegiance after expelling Soviet advisers in the late 1970s. The U.S. provided significant aid to Somalia until 1989.) After Desert Storm, Gonzo Station was of less strategic importance as it was too far away to respond to operational needs in Iraq and the Persian Gulf and it is currently used as a "transit area." The Indian Ocean operations were supported by facilities at Diego Garcia, an atoll about 1,800 nautical miles east of the coast of Africa, 1,200 nautical miles south of the southern tip of India and 3,467 nautical miles west of Darwin, Australia. The Navy operates a large naval ship and submarine support base, military air base, communications and space tracking facilities, and an anchorage for pre-positioned military supplies for regional operations on Military Sealift Command ships in the lagoon. (Diego Garcia is part of the British Indian Ocean Territories and under an agreement with the U.S. was built up in the 1970s to allow an American presence in the Indian Ocean and make up for the loss of bases in Vietnam and the Philippines.)

The Reagan Buildup

After the end of the Vietnam War, all the military services had seen reductions in their budgets, while the Soviet Union, which had supported North Vietnam, began operating from ports in South Vietnam. The Soviet navy also began to operate globally, even into the Gulf of Mexico, while Soviet ground and air forces stepped up their deployments in Eastern Europe. Against this backdrop, a key element of the 1980 Reagan presidential campaign had been a pledge to restore America's strategic capabilities with the development of new weapons, such as the B-1 bomber, the Abrams tank, and the Bradley fighting vehicle. The Navy was the biggest winner under the Reagan administration with Navy Secretary John Lehman advocating a "600 ship" Navy with new weapons systems such as the *Ohio*-class ballistic missile submarine armed with the Trident nuclear missile, *Los Angeles*–class attack submarines, and *Ticonderoga*-class guided-missile cruisers with revolutionary new

Aegis combat systems. The *Iowa*-class battleships of World War II vintage were also re-commissioned with modern weapons, such as the RGM-84 Harpoon, BGM-109 Tomahawk, and Phalanx CIWS systems, while the construction of *Nimitz*-class nuclear carriers was stepped up. Several carriers were also modernized under Service Life Extension Programs (SLEPs) to add 15 years to their useful service lives. The F/A-18 Hornet entered service along with improved versions of the F-14 Tomcat, the EA-6B Prowler, and the A-6 Intruder. The centerpiece of the new approach was the "Lehman Doctrine," a plan that called for a military response to any Russian invasion in Europe by attacking and invading the Soviet Far East along the Pacific, a much less defended front. American forces would then sever the Trans-Siberian Railroad and fight their way westward toward Moscow.[5] The Soviets, for their part, had other things to deal with as they became embroiled in Afghanistan from 1979 until 1989.

Libya

Relations between the radical regime of Libya's Muammar Gaddafi and the U.S. had been strained since the early 1970s. In 1973 he declared the Gulf of Sidra to be part of Libyan territorial waters and the U.S. responded by conducting naval operations in the Gulf to assert its freedom of navigation rights under international law. Gaddafi declared that crossing the "line of death" would invite a military response and Libyan aircraft often confronted U.S. naval forces in and near the Gulf. On two occasions its fighters opened fire on U.S. reconnaissance flights off the Libyan coast: once in the spring of 1973 and again in the fall of 1980. When Ronald Reagan came to office in January 1981, freedom of navigation operations were stepped up. In August 1981 Reagan authorized a naval force, led by the *Forrestal* and *Nimitz*, to deploy into the disputed area to conduct an exercise. In response, the Libyan Air Force deployed several interceptors and fighter-bombers. Early on the morning of 18 August, when the exercise began, at least three MiG-25 Foxbats approached the carrier groups, but were escorted away by F-4 Phantoms from *Forrestal* and F-14s from *Nimitz*. Thirty-five pairs of MiG-23 Floggers, MiG-25s, Sukhoi Su-20 Fitter-Cs, Su-22M Fitter-Js and Mirage F1s flew into the area to locate the task groups, but were soon intercepted by seven pairs of F-14s and F-4s. (A later assessment indicated that a MiG-25 may have fired a missile from 18 miles away at U.S. fighter aircraft that day.) The next day, 19 August, two Libyan Su-22 Fitter attack aircraft fired upon and were subsequently shot down by two *Nimitz* F-14 Tomcats off of the Libyan coast.

After the hijacking of TWA Flight 847 in June 1985 and the Rome and Vienna airport attacks in December of that year, the U.S. claimed Gaddafi was involved through his support of Palestinian terrorist Abu Nidal. At the same time Libya began installing Soviet-supplied SA-5 Gammon missile batteries and radars.[6] The first two freedom of navigation exercises in January and February 1986 were without incident. The third began on 23 March with the *America*, *Coral Sea*, and *Saratoga* accompanied by five cruisers, twelve destroyers, and six frigates. The next morning the cruiser *Ticonderoga*, accompanied by destroyers *Scott* and *Caron*, covered by fighters, crossed the Line of Death. A Libyan missile site near Sirte launched two SA-5 SAMs toward the F-14A Tomcats of *America's* VF-102. The missiles missed and fell harmlessly into the sea. Two more SA-5s were launched but were jammed by an EA-6B Prowler. Two hours later, two MiG-23s took off from Benina air base with orders to intercept and shoot down some of the U.S. fighters. Before the Libyan aircraft could get close enough, an E-2C Hawkeye detected them and alerted two F-14s from VF-33, which intercepted the MiGs at 20,000 feet. The Libyans began maneuvering to get into firing positions on the two F-14s—a clear sign of hostile intent. *America* gave the pilots approval to open fire if necessary. An intense dogfight ensued, though without any missiles being fired. The F-14s dropped to 5,000 feet and put themselves between the sun and the Libyans, maneuvering into a six o'clock position behind the MiGs. They were ready to shoot when the MiGs moved off toward their base. One of them turned back into the F-14s, but again turned away and headed south. In the meantime several Libyan patrol boats headed out toward the U.S. battle group, and aircraft were sent out to counter them. The *Saratoga* launched A-7s armed with HARM missiles from the VA-83 Rampagers, A-6s armed with Harpoon missiles and cluster bombs from the VA-85 Black Falcons, and EA-6Bs from the Scorpions of VAQ-132. *America* had A-6s from the VA-34 Blue Blasters and EA-6Bs from the Marine squadron VMAQ-2 and *Coral Sea* had A-6s from the VA-55 Warhorses and EA-6Bs from VAQ-135's Black Ravens in the air, supported by several E-2Cs, F-14s, F/A-18s, and KA-6Ds. The first air strikes occurred around 1926 when two A-6 Intruders from VA-34 found a *Beir Grassa*–class patrol boat (a French-built version of the *La Combattante IIa*); the ship was first disabled by a Harpoon missile fired by one of the A-6 Intruders from VA-34 (the first combat use of the Harpoon) and then destroyed by Intruders from VA-85 using Rockeye cluster bombs.[7] Forty minutes later, F-14s, F/A-18s, A-7Es, and EA-6Bs headed toward the SA-5 site near Sirte at low level and suddenly climbed, causing the Libyans to activate their radars and launch missiles. In reply, the A-7s launched several HARMs. The strike formation then dropped down to above sea level and turned back to their carriers. Meanwhile, A-6s from VA-86 and VA-55 engaged several Libyan missile boats. At around 2155 two A-6s from VA-55 attacked a *Nanuchka*-class corvette headed toward the cruiser *Yorktown*, prompting *Richmond K. Turner* to open fire with a Harpoon missile that struck the Libyan vessel and started a fire. (The corvette survived the attack and was towed back to Benghazi.) Around midnight, the Libyans launched several SA-2s and SA-5s; this time at the A-6s and A-7s responded by heading toward the coast. A-7s from VA-83 launched HARM missiles, disabling several Libyan radars. Three more SA-5s were launched from Sirte with a single SA-2 launched near Benghazi. At 0730 another Libyan corvette was intercepted by A-6s from VA-55 and was disabled by Rockeye (it was later sunk by a Harpoon missile from a VA-85 A-6). The operation, which had been given the name Prairie Fire, ended after this strike with no losses.[8]

In response to the 5 April 1986 terrorist bombing carried out by Libyan agents of a nightclub in West Berlin, in which three people were killed and 229 others injured, Air Force, Navy, and Marine aircraft conducted strikes against terrorist centers in Libya. This operation, called El Dorado Canyon, was carried out on 15 April 1986. The operation started on 14 April, when Air Force F-111s departed England on a 3,500 mile flight with four aerial refuelings each way (due to overflight restrictions by France, Spain and Italy). As the aircraft approached Libya, the *Saratoga*, *Coral Sea*, and *America* launched 14 A-6E strike aircraft and 12 F/A-18 and A-7 strike support aircraft. As EF-111 Raven aircraft jammed Libyan air defenses, Navy support aircraft suppressed SAM sites. Attacking Navy aircraft hit Benina Airfield and the Benghazi military barracks, while Air Force F-111s struck the Aziziyah barracks in Tripoli and the Sidi Bilal terrorist training camp. The last strike was by Air Force F-111s against the Tripoli military airport. All the Navy aircraft had been recovered safely, but one F-111 had been lost. The attack left 37 dead and 93 injured; Gaddafi had been warned at the last minute and had escaped injury.

The next Gulf of Sidra incident occurred in 1989. After the U.S. accused Libya of building a chemical weapons plant near Rabat, the *John F. Kennedy* deployed near the Libyan coast. (A second carrier group with the *Theodore Roosevelt* was also ready to sail into the Gulf of Sidra.) On the morning of 4 January 1989 the *Kennedy* battle group was operating some 70 miles north of Libya, with a group of A-6 Intruders on an exercise south of Crete, escorted by two pairs of F-14As from the VF-14 Tophatters and VF-32 Swordsmen, and as well as an E-2C from the VAW-126 Seahawks. Later that morning the southernmost Combat Air Patrol (CAP) station was taken by two F-14s from VF-32 (Commander Joseph Bernard Connelly and Commander Leo F. Enwright; and Lieutenant Herman C. Cook III and Lieutenant Commander Steven Patrick Collins). The crews had been briefed to expect some kind of hostilities. At 1150 the E-2 informed the F-14s that four Libyan MiG-23s had taken off from Al Bumbaw airfield, near Tobruk. The F-14s turned toward the first two MiG-23s some 30 miles ahead of the second pair and acquired them on radar. At the time the Floggers were 72 miles away at 10,000 feet and heading directly toward the Tomcats and the carrier. The F-14s turned away from the head-on approach to indicate that they were not attempting to engage. The Floggers changed course to intercept. The F-14s descended to 3,000 feet to give them a clear radar picture of the Floggers against the sky and leave the Floggers with sea clutter to contend with. Four more times the F-14s turned away from the approaching MiGs. Each time the Libyan aircraft turned in and continued to close. At 1159 the lead Tomcat armed its Sidewinder and Sparrow missiles. The E-2 had given the F-14 crews authority to fire if threatened—they did not have to wait until after the Libyans opened fire. At 1201 the lead Tomcat fired a Sparrow at a range of 14 miles. It failed to track because of a wrong switch setting. At ten miles a second Sparrow missile was launched, but it also failed to track its target. The Floggers accelerated and continued to approach. At six miles the Tomcats split and the Floggers followed the wingman while the lead Tomcat circled to get on their tails. The wingman fired a third Sparrow from five miles, downing one of the Floggers. The lead Tomcat had by now closed in behind the other Flogger and closed to within a mile and a half before firing a Sidewinder, which hit its target. The Tomcats headed north to return to the *Kennedy*. Although the Libyan pilots were both seen to eject and parachute into the sea, they were never recovered.[9]

Lebanon

In June 1982 Israel invaded Lebanon to once again attempt to stop attacks into its territory launched from its neighbor. In August, in a move intended to separate the combatants and establish peace and order, several countries deployed their troops to Lebanon. U.S., French, and Italian units started arriving in Beirut, establishing what later became known as the Multi-National Force (MNF). After some initial clashes with Syrian troops, who controlled much of Lebanese territory (the Syrians had moved into areas formerly occupied by Israeli forces as they withdrew according to the initial peace-keeping agreements), the situation remained relatively stable until April 1983. Unfortunately, some factions within Lebanon began to consider the presence of the MNF troops as interference on the part of Western powers. The Israelis saw Western intervention as support for their efforts and interests; Muslims believed the MNF was there to support the Christians and protect the Israelis; and the Christians thought the MNF would help increase their influence in the country. Eventually, the MNF came to be regarded as one of the contending factions and not an impartial peace-keeping force. As the Marines at their compound at the Beirut International Airport came under attack, they responded with air strikes and offshore gunfire support from the battleship *New Jersey*, which had arrived from the Caribbean. The *Eisenhower*, which had been operating off the coast of Libya, was also sent to an area east of Cyprus named Bagel Station. On 23 October two suicide truck bombs struck the buildings housing U.S. and French forces, killing 299. (U.S. casualties included 220 Marines, 18 sailors, and 3 soldiers. French casualties included 58 paratroopers. There were also a few Lebanese civilians killed.) Islamic Jihad, later known as Hezbollah, claimed responsibility for the bombing.

The situation came to a head when the Israelis struck Syrian SAM sites in Lebanon on 16 November. While there were casualties on the Syrian side, one of the attacking Israeli Kfirs was also shot down. The Israelis struck again on 3 December. At the same time as the Israeli strike, two F-14s of VF-32 from the *Kennedy* were in the area on a reconnaissance mission (one of the F-14s was equipped with a TARPS pod) and were fired on by several different SAMs, eventually forcing them to abort their mission. The U.S., however, saw this as a clear provocation and, having a clear target to hit back at, ordered Rear Admiral Jerry Tuttle, commander of Task Force 60, to begin planning a retaliatory strike that evening. Admiral Tuttle had left the planning to teams on the *Independence* and *John F. Kennedy* (the *Eisenhower* had been relieved at the end of November). The strikes were to come in a low altitude at 1100 the following day. Inexplicably, orders from Washington directed that the strikes be made at 0545 and gave very specific instructions about the targets to be attacked, weapons used, as well as the altitude—20,000 feet! Moving the attack time up resulted in chaos on the carriers as there was not enough time to swap out ordnance, resulting in haphazard weapons loads; worse, the support aircraft that should have been launched first to be in position before the strike aircraft went "feet dry" (in naval aviation parlance this means flying over land) were launched after the first strike aircraft. Although the two VF-31 F-14s caught up with the strike, the E-2 was late getting into position and could not effectively control the strikes. Two EA-6Bs were behind the strike force and could only react to events as they occurred instead of proactively jamming potential threats. Almost immediately, the attackers were met with several incoming Syrian SAMs, forcing the formations to break up and take evasive action. Within seconds an A-7 from VA-15 was hit and her pilot headed out to sea ("feet wet") to eject, to be picked up by helo. With the breakup of the formation, aircraft attacked individually instead of being able to provide mutual support. As the Intruders rolled in on their targets one by one, an A-6 from VA-85 was hit by a man-portable missile as it pulled out of its dive at low altitude. The pilot, Lieutenant Mark "Doppler" Lange, did his best to keep the aircraft airborne so his bombardier-navigator (BN), Lieutenant Bobby Goodman, could eject safely. It crashed on a hill, directly above a village surrounded by Syrian AAA-positions. Lieutenant Lange ejected at the last moment, but his parachute had not deployed properly by the time he hit the ground. Lange died of his injuries shortly afterwards while in the hands of Syrian troops and Lebanese civilians. Goodman broke three ribs and injured his shoulder and knee during the landing. He was captured by the Syrians and taken to Damascus. (Lieutenant Goodman was eventually repatriated, but not before he was paraded by the Syrians for world media.) Commander Andrews, CAG CVW-6, started searching for the crew of the downed Intruder, though the chances of recovering them were slim. Upon reaching the area where the A-6 had gone down, he circled until the Syrian flak opened up. He attacked the positions he could identify with 20mm cannon fire, but on his last pass his A-7 took a direct hit from a SA-7. Commander Andrews went "feet wet" near Beirut, where he ejected safely. Although two SAR helicopters were dispatched to the area, Andrews was picked up by a local fisherman and then handed over to the Marines. The strikes were claimed to be successful, but the situation in Lebanon had deteriorated. Although the *New Jersey* continued to support the Marines, the fate of the MNF was sealed. By February 1984 the U.S., French, Italian, and British troops were forced to withdraw from Lebanon, leaving the country in the chaos of civil war.[10]

Urgent Fury

In 1983 the U.S. invaded Grenada, a Caribbean island nation with a population of about 90,000 located 100 miles north of Venezuela. Grenada had gained independence from the U.K. in 1974 and a leftist movement seized power in a coup in 1979, suspending the constitution. An internal power struggle in 1983 led to the death of revolutionary Prime Minister Maurice Bishop. On the justification of acting to protect or evacuate American citizens, as well as to provide stability in the region and promote the establishment of a more democratic government, American forces launched Operation Urgent Fury on 25 October 1983. (This was less than 48 hours after the bombing of the U.S. Marine barracks in Beirut.) About 7,000 American troops were involved. The 1st and 2nd Ranger Battalions along with paratroopers from the 82nd Airborne Division formed the U.S. Army's Rapid Deployment Force. There were also Marines, Army Delta Force, and Navy SEALs and a few hundred other troops from the Caribbean Peace Forces. The Rangers overcame Grenadian resistance after a low-altitude airborne assault on Point Salines Airport on the southern end of the island while a Marine helicopter and amphibious landing took place on the northern end at Pearl's Airfield shortly afterward. During the landings, aircraft from the *Independence* flew missions in support of the operation. The island was secured after several days of fighting. (Although the landings had broad support from the American public, there was international criticism. The operation, as the first combat operation since Vietnam, also highlighted problems with interoperability among the services, which eventually led to the passage of the Goldwater-Nichols Act in 1986.)[11]

Achille Lauro

On 7 October 1985 four Palestine Liberation Front (PLF) terrorists hijacked the Italian liner MS *Achille Lauro* off Egypt as she sailed from Alexandria to Port Said. The next day, after the Syrian government refused to let the ship dock at Tartus, the hijackers singled out Leon Klinghoffer, an elderly Jewish man in a wheelchair, shooting him in the forehead and chest and dumping his body and wheelchair overboard. After two days of negotiations, the hijackers agreed to abandon the liner in exchange for safe conduct and were flown toward Tunisia on an Egyptian commercial airliner. President Reagan, who had the Navy's SEAL Team Six and Delta Force standing by for a possible rescue attempt, ordered the intercept of the airliner. On 10 October F-14s from *Saratoga*'s VF-74 Be-Devilers and VF-103 Sluggers intercepted the airliner and directed it to land at Naval Air Station Sigonella, a NATO base in Sicily, where the hijackers were arrested by the Italians. (There was a disagreement between American and Italian authorities that had taken hours to resolve. The other passengers on the plane were allowed to continue on to their destination, despite protests by the U.S. government that the mastermind of the hijacking, Muhammad Zaidan, was among them. For its part, the Egyptian government demanded an apology from the U.S. for diverting the flight.)[12]

The Tanker Wars

In 1980 Iraq had invaded neighboring Iran, hoping to take advantage of the disorder following the Iranian revolution to settle long simmering border disputes and to eliminate the influence of radical Shi'a Islam on Iraq's suppressed Shi'a majority. After initial Iraqi advances, the Iranians regained their lost territory and went on the offensive. Iran had blocked Iraqi exports of oil via the Shatt-al-Arab waterway, while Syria had closed Iraq's pipeline to the Mediterranean in an attempt to strangle the Iraqi war effort through its economy. Although the Iraqi government under Saddam Hussein was disliked by other Arab states, they feared Iranian Islamic fundamentalism more. Jordan opened Aqaba to Iraqi imports and new pipelines were laid to the Red Sea and Turkey. Iraqi exports also went through Kuwait. (The Arab states also directly funded Iraq to the sum of about $60 billion.) In 1984 Iraq attacked Iranian tankers

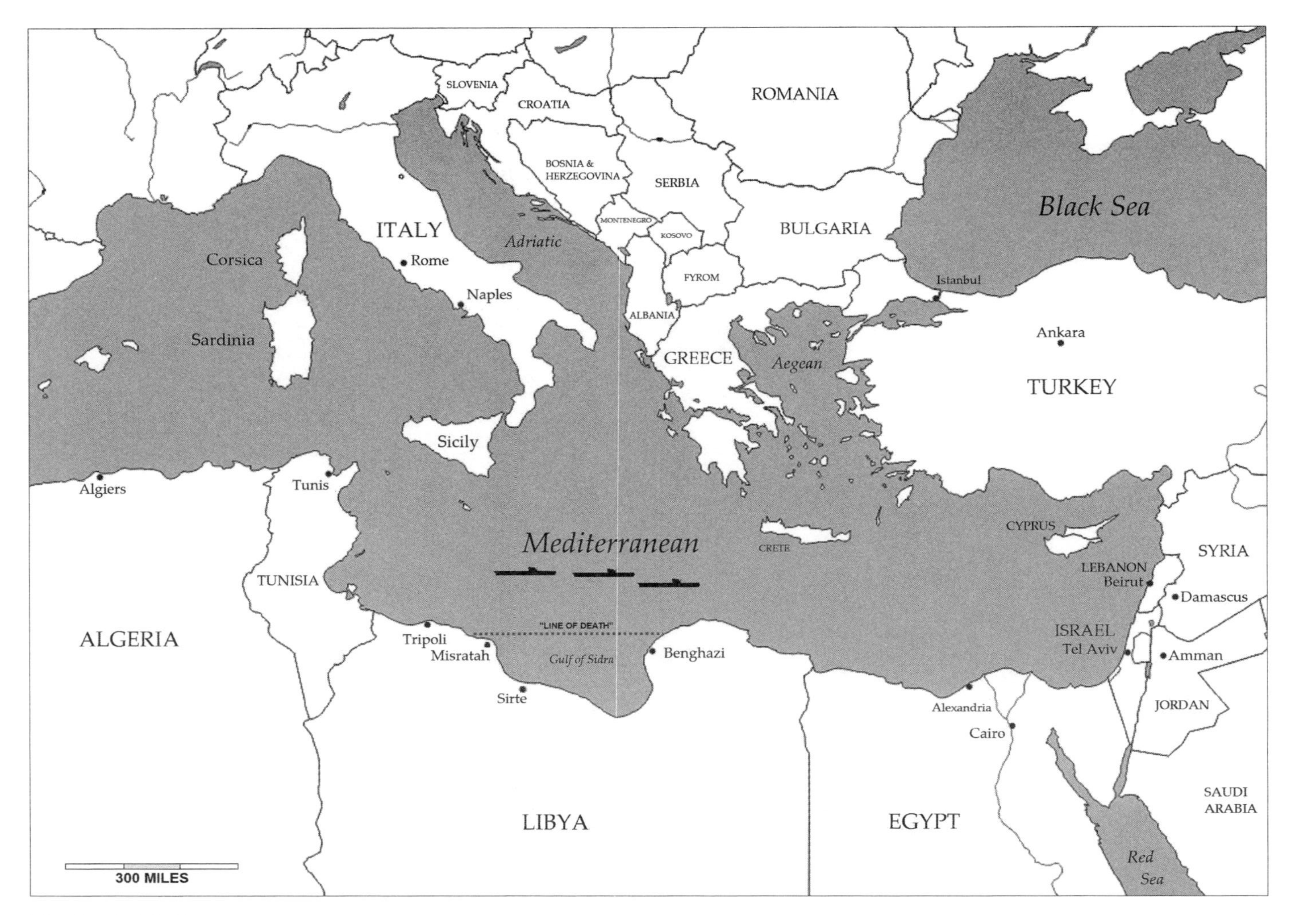

Map 5. The Mediterranean

Map adapted by the author from http://d-maps.com/carte.php?num_car=33696&lang=en.

and the vital oil terminal at Kharg Island. Iran struck back by attacking tankers carrying Iraqi oil from Kuwait and later any tanker of the Gulf states supporting Iraq. (The air and small boat attacks did little to damage the economies of either country. The price of oil was never seriously affected and Iran just moved its shipping port to Larak Island in the straits of Hormuz.) In December 1986 Kuwait asked the U.S. to offer protection to its tanker fleet and U.S. warships would subsequently begin patrolling the Gulf in 1987. (Since U.S. law forbids the use of Navy ships to escort civilian vessels under a foreign flag, the Kuwaiti ships were re-registered under the U.S. flag.)

On 17 May 1987 an Iraqi Super-Etendard aircraft fired two Exocet missiles at the frigate *Stark*, mistaking her for an Iranian warship. The *Stark*'s radar had not detected the incoming missiles and warning was given by a lookout only moments before they struck. The first Exocet failed to detonate when it penetrated the port side, but left flaming rocket fuel in its path. The second struck at almost the same point, and exploded in crew quarters, killing 37 sailors and injuring 21. On fire and listing, with a 13-foot-wide gash in her hull, her crew brought her under control during the night as she made her way to Bahrain for temporary repairs before returning to her home port in Mayport, Florida, under her own power. (The *Stark* was eventually repaired at Ingalls Shipbuilding in Mississippi. Iraq apologized for the attack and to prevent further incidents, U.S., Iraqi and Saudi forces began collaborating.)

In July 1987 the U.S. Navy began escorting the re-flagged Kuwaiti tankers as part of Operation Earnest Will. In what became the largest naval convoy operation since World War II, the Seventh Fleet (which had primary responsibility for combat operations in the Persian Gulf region) employed ships from the Pacific's Third and Seventh Fleets and the Mediterranean-based Sixth Fleet. These ships usually operated in and near the Gulf during parts of their normal six-month deployments.

On 16 October 1987 a reflagged Kuwaiti oil tanker was at anchor off Kuwait when it was hit by an Iranian Silkworm missile. In response, the U.S. decided to attack two platforms in the Reshadat oil field that had been used by the Iranian Revolutionary Guard Corps. Under the code name Nimble Archer, the *Ranger* provided air cover by two F-14 Tomcats and an E-2 Hawkeye. The platform crews were warned to abandon them 20 minutes before a surface action group opened fire.[13]

Iran accused the U.S. of helping Iraq and began to sow the Gulf with anti-ship mines. Several ships were hit and on 14 April 1988 the frigate *Samuel B. Roberts* hit a mine while operating in the central Persian Gulf, an area it had safely transited a few days previously. The mine blew a 15-foot hole in the hull, flooded the engine room, knocked the two gas turbines from their mounts, and broke the keel of the ship; fighting fire and flooding for five hours, her crew managed to save their ship. Tensions rose even further when the Aegis guided-missile cruiser *Vincennes* shot down Iran Air Flight 655 as it flew over the Strait of Hormuz, killing all 290 civilians on board, on 3 July.

A Dangerous World 13

The Gulf War

When the Iran-Iraq War, which resembled World War I in its trench warfare and bloodshed, ended with a cease-fire in 1988, both countries had suffered grievously, both economically and in loss of life—perhaps as many as half a million Iraqi and Iranian soldiers and civilians had died. Instead of rebuilding his country, Iraq's Saddam Hussein threatened neighboring Kuwait and the United Arab Emirates, two other countries that Iraq held ancient claims to, to forgive Iraq's war debts, pay compensation for Iraq's "protection" from Iran, and cut oil production to raise oil prices. Although they did agree to cut oil production, the demands were just a pretext for invasion. In the hours before dawn on 2 August 1990, Iraqi tanks and armored units rolled down the superhighway connecting the two countries, covered by Mi-24 Hind helicopter gunships and Sukhoi ground attack aircraft. The Kuwaitis fought back with U.S.-built Hawk surface-to-air missiles (SAMs), but quickly exhausted their supply of missiles. Kuwaiti pilots tried to intervene, but quickly found it better to fly to safety in Saudi Arabia. By dawn Iraq had control of Kuwait.

That day the U.N. Security Council demanded that Iraq withdraw immediately. Fearing his country would be next, King Fahd of Saudi Arabia asked for help, offering air bases and facilities. The United States began to mobilize by sending 48 F-15 Eagles to the area. On 8 August Hussein responded by annexing Kuwait. Later that same day President George H.W. Bush announced he was sending troops to the region. The six ships of the Middle East Force had been on station in the Persian Gulf when the President decided to deploy additional forces to Southwest Asia. The *Independence* battle group had steamed into strike range in the North Arabian Sea from the Indian Ocean, and the *Eisenhower* battle group transited the Suez Canal from the Eastern Mediterranean to be within strike range of Iraq from the Red Sea. When U.N. sanctions were imposed, the U.S. Navy immediately began to intercept any ships headed to or from Iraq and Kuwait. Although the United States provided the bulk of the coalition forces involved, Great Britain, France, Argentina, Belgium, Egypt, Germany, Italy, Canada, New Zealand, South Korea, Bahrain, Qatar, the United Arab Emirates, Greece, the Netherlands, and Australia all sent either aviation or naval forces. (On 2 August U.N. Resolution 665 was passed, giving the Coalition legal status.) Operation Desert Shield had begun.

The initial focus of the Coalition had been to protect Saudi Arabia from a possible invasion and the presence of two carrier battle groups within striking distance were probably a factor in deterring Saddam Hussein from the attempt. By November the focus shifted to expelling Iraq from Kuwait. On 29 November the U.N. passed Resolution 678, ordering Iraq to leave by 15 January 15 1991. The U.S. Congress authorized President Bush to use force against Iraq on 12 January 1991. The stage was set for one of the largest joint and combined military operations in the 20th century.[1]

On 17 January 1991, soon after midnight local time, Air Force F-117A Nighthawk stealth fighters flew into Baghdad, dropping laser-guided bombs on various sites around the city while the Navy launched Tomahawk missiles from the Red Sea against other key targets. Ultimately, six aircraft carrier battle groups and two battleships (*Wisconsin* and *Missouri*) were involved in the campaign, while cruisers, destroyers, battleships, and submarines launched a combined total of 288 Tomahawk cruise missiles from the Red Sea and Persian Gulf against heavily defended key Iraqi facilities, contributing to the neutralization of Iraqi air defenses. About one third of all strike sorties were flown by Navy and Marine Corps aircraft.[2] *America* and *Theodore Roosevelt* had left Norfolk on 28 December 1990, joining *Midway*, *Saratoga*, *John F. Kennedy*, and *Ranger*, who were already on station, just in time for the opening of Desert Storm. (The *Eisenhower* had departed early in September and the *Independence* was relieved by the *Midway* in November.[3]) The first priority for Coalition forces was to destroy the Iraqi air force and anti-aircraft defenses. The sorties were launched mostly from Saudi Arabia and the six Carrier Battle Groups (CVBGs) in the Persian Gulf and Red Sea. Next, the Coalition targeted command and communication facilities. Since Saddam Hussein had closely micromanaged the Iraqi forces in the Iran-Iraq War, and initiative at lower levels was discouraged, Coalition planners hoped that Iraqi resistance would quickly collapse if the national leadership was "decapitated," that is, deprived of command and control. The third and largest phase was against military targets throughout Iraq and Kuwait. Within five hours after the first attacks, Iraqi state radio broadcast a voice claimed to be that of Saddam Hussein declaring that "The great duel, the mother of all battles has begun. The dawn of victory nears as this great showdown begins."[4] Saddam Hussein had promised that if Iraq was invaded, he would attack Israel and hoped that Israeli involvement would disrupt the Coalition. Iraq launched eight Al Hussein missiles at Israel the next day. (The Al Hussein was an Iraqi-built upgraded version of the Soviet-made Scud missile with longer range. All of the attacks made during the war were made by the Al Hussein, but the term "Scud" was commonly used in most news reporting.) But

Top: The destroyer *Rupertus* maneuvers to within 20 feet of the *Forrestal* so fire hoses could be used effectively, 29 July 1967. More than 130 crew were killed in the blaze off the coast of Vietnam.

Right: Ordnance continues to explode as smoke billows from the aft flight deck of the *Forrestal*.

Bottom: Flames threaten to engulf the island of the *Forrestal* as the fire rages.

Top: Wreckage of destroyed aircraft on the *Forrestal*'s flight deck.

Right: Firefighting hoses from the *Rupertus* spray water on the fire as she comes alongside the *Forrestal*'s port quarter. Excess water from the firefighting can be seen draining through openings in the carrier's side.

Bottom: Air wing and ship's company volunteered when many of the *Forrestal*'s trained firefighters were killed in the initial blasts.

A stern view of the *John F. Kennedy* under way, June 1982.

F-14, F/A-18, A-6, E-2 EA-6B, S-3, and C-2 aircraft from CVW-15 fly over the *Kitty Hawk*, 1993.

Left: *Forrestal* crewmen spell out "We Love N.Y." during Fleet Week, April 1989.

Bottom: A view of the *Forrestal*'s island, March 1989.

Left: A plane director on board the *Saratoga* directs an F-14 Tomcat into position on the starboard catapult, 1986.

Center: The nuclear-powered *Enterprise* steams alongside the *Forrestal*, 1988.

Bottom: Aircraft movements on the *Saratoga* are tracked in flight control using the "Ouija" board, 1986.

An F-14 Tomcat from VF-2 flies past the *Ranger* while she operates off southern California, January 1989.

Crewmen on the *Ranger* man the rail as she returns to North Island after deploying to the Persian Gulf during Operation Desert Storm, June 1991.

The *John F. Kennedy* fires a NATO Sea Sparrow as part of a missile exercise during her deployment in support of Operation Enduring Freedom, July 2002.

The *John F. Kennedy* conducts a Phalanx Close-In Weapons System (CIWS) live fire exercise while conducting proficiency training off the Florida coast, April 2006.

The *Kitty Hawk* tests her newly installed Rolling Airframe Missile (RAM) while working up for deployment, March 2002. Only the *Kitty Hawk* and *Kennedy* were equipped with RAM systems.

The men of the *Ranger* spell out the reason for operating off Somalia—supporting Operation Restore Hope, January 2000.

Two restored World War II B-25 Mitchell bombers on the flight deck of the *Ranger*. The B-25s were launched in a re-enactment of the Doolittle Raid of 1942, April 1991.

When the *Ranger* returned to North Island after deploying to the Persian Gulf during Operation Desert Storm, she was greeted by her mascot, the Lone Ranger, June 1991.

A bow view of the *Saratoga* under way during Operation Desert Storm, February 1991.

Two F/A-18 Hornets from VFA-74 fly over the *Saratoga* during Operation Desert Shield, November 1990.

A port beam view of the *Saratoga* under way during Operation Desert Shield, December 1990.

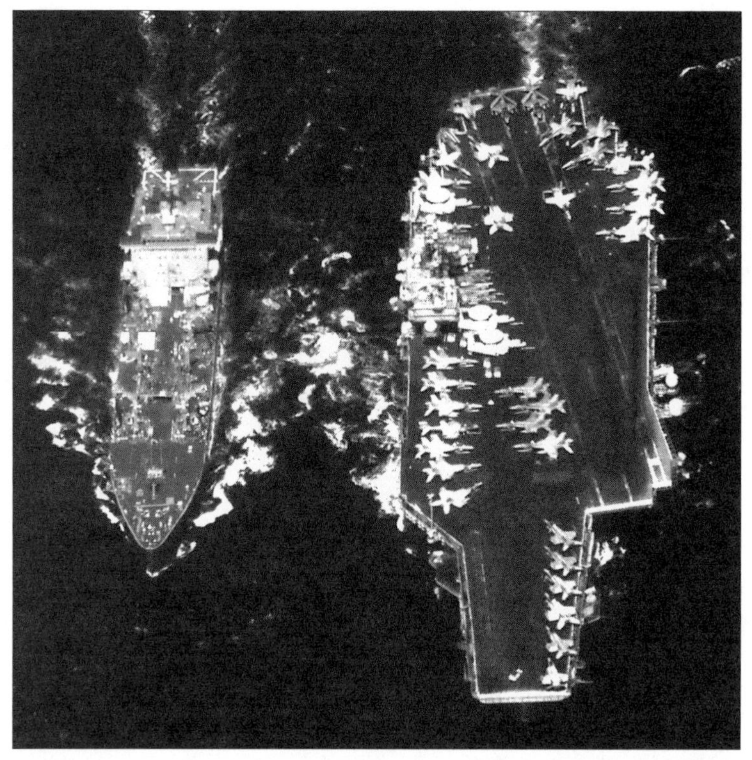

The *Kitty Hawk* conducts a replenishment at sea with the Military Sealift Command fleet replenishment oiler USNS *Guadalupe*, July 2008. (This picture was taken by a RQ-4A Global Hawk unmanned aerial vehicle [UAV] surveillance aircraft.)

An F/A-18 Hornet from VFA-195 launches from the *Kitty Hawk* while she was under way in the Pacific for the last time, August 2008. CVW-5 left to join the *George Washington*, which replaced her as the forward-deployed carrier in Yokosuka, Japan, in September.

Right: "Blue Shirt" aircraft handlers on board the *John F. Kennedy* watch flight operations from "Vultures Row." New flight deck personnel were required to observe flight operations for three days before they could work on the flight deck. She was conducting Carrier Qualifications off the Cherry Point operating area at the time, November 2003.

Center: The *Independence* under way off the coast of Lebanon while supporting Marine peace-keepers in Beirut, December 1983.

Bottom: The *Independence* is dressed out in bunting for her decommissioning ceremony in Bremerton, September 1998.

Top: The *Constellation* departs North Island at the start of a six-month deployment to support Operation Enduring Freedom, November 2002.

Right: An F/A-18 Hornet aircraft assigned to VMFA-314 prepares for launch from the *Constellation* during an operational evaluation (OPEVAL), February 1984.

Bottom: The *Constellation* displays a sign with the famous quote of the late Todd Beamer "Okay, Let's Roll" as she departs for a six-month deployment to the Western Pacific, November 2002.

The *Constellation* and *Kitty Hawk* steam alongside one another while deployed in support of Operation Iraqi Freedom, April 2003.

Left: An F-14 Tomcat from VF-2 prepares to launch from the *Constellation* while supporting Operation Enduring Freedom, January 2003. The squadron was decommissioned along with the *Constellation* after this deployment.

Bottom: Ordnance is staged in the *Constellation*'s hangar bay before being loaded onto aircraft on the flight deck, March 2003. The *Constellation* was deployed in support of Operation Iraqi Freedom at the time.

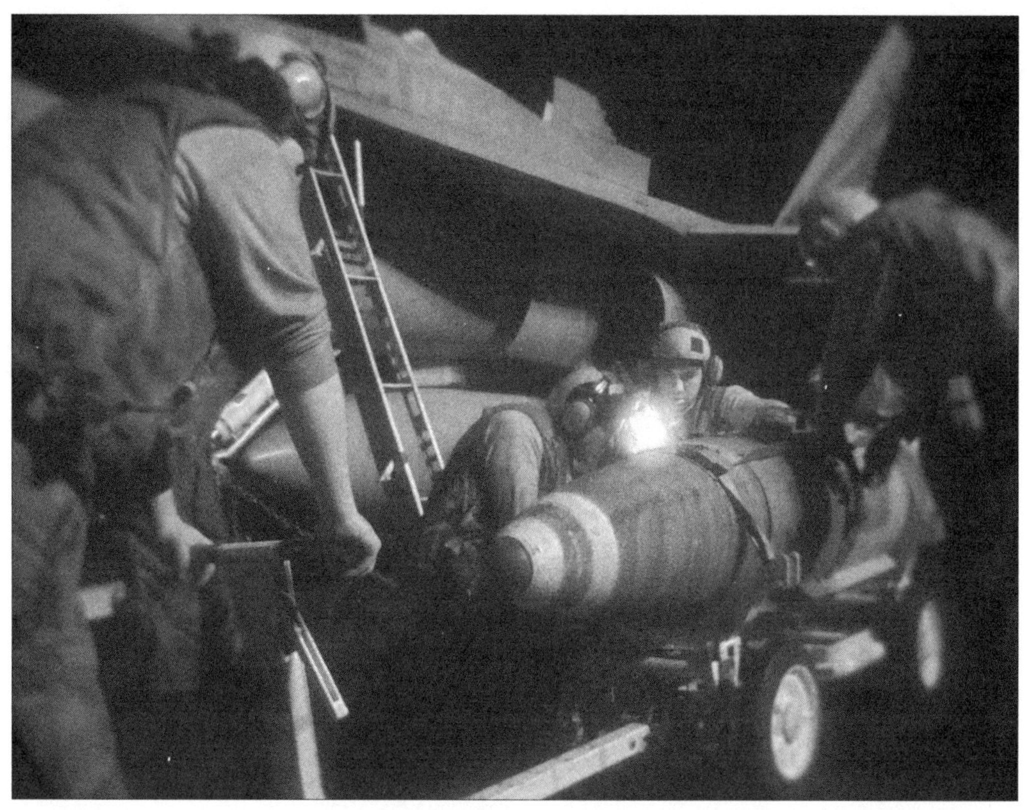

Aviation Ordnancemen load GBU-31 Joint Direct Attack Munitions (JDAM) onto an F/A-18 Hornet on the flight deck of the *Constellation* during Operation Iraqi Freedom, March 2003.

The decommissioned *Forrestal* arrives at Naval Support Activity Philadelphia from Naval Station Newport, June 2010. (Once decommissioned, she was officially referred to as the "Ex-USS *Forrestal*.")

The *John F. Kennedy* arrives at Pensacola after completing a training exercise in the Gulf of Mexico, March 2004.

Distinguished visitors and guests attend the decommissioning ceremony of the *John F. Kennedy* in Mayport, March 2007.

Top: Aerial view of sailors spelling out Sayonara on the flight deck for the *Kitty Hawk* as she departs Yokosuka, Japan, May 2008.

Center: Korean, Japanese, Chilean, and Australian ships steam in formation with the *Kitty Hawk* during a Rim of the Pacific (RIMPAC) exercise off Hawaii, July 2008.

Right: Approximately 250 sailors on board *Kitty Hawk* display a 1,065-foot homecoming pennant on the flight deck as the ship approaches North Island, August 2008. She was returning to San Diego for the first time in ten years.

President Bush had prevailed on the Israelis not to respond militarily and rushed Patriot surface-to-air missiles to defend against the Scuds.[5] The missile attacks on Israel continued throughout the war, with a total of 42 Scuds fired by Iraq into Israel during the next seven weeks of the war. While Israeli casualties were relatively light, there was extensive property damage and widespread fear that the Iraqis would use chemical weapons against the civilian population. About a third of the Coalition air strikes were devoted to attacking Scuds, some of which were on trucks and hard to locate. (Also, some U.S. and British special forces teams had been covertly inserted into western Iraq on search and destroy missions against the Scuds.) Saudi Arabia was also targeted with 47 Scuds and one was fired at Bahrain and another at Qatar. (These missiles were fired at both military and civilian targets.)

Within two weeks of the start of the air campaign, Iraq engaged in what was described as "an act of environmental terrorism" by pumping oil through an underground pipeline from Kuwait into the Arabian Gulf. The slick from the oil would grow to over 35 miles and threaten to foul Saudi desalination plants. (The flow was stopped when Air Force F-111 strikes used laser-guided bombs to destroy the pipes and ignite the oil and burn off pollutants.)

On 29 January Iraqi forces attacked and occupied the lightly defended Saudi city of Khafji with tanks and infantry. The Battle of Khafji ended two days later when the Iraqis were driven back by the Saudi National Guard and the Marines, supported by Qatari forces with extensive allied close air support and artillery fire. Khafji was a strategically important city immediately after the Iraqi invasion of Kuwait. Iraqi reluctance to commit several armored divisions to occupy Khafji may have been related to Iraqi fears of a Marine amphibious landing on the Kuwaiti shore. The threat of such a landing kept several Iraqi units tied down to repel an invasion that never came. American feints by air attacks and naval gunfire the night before the liberation of Kuwait were designed to make the Iraqis believe the main coalition ground attack would focus on Central Kuwait.[6]

In the early morning hours of 24 February, local time, the ground war began with Army, Marine, and Arab forces moving into Iraq and Kuwait when the Iraqis failed to meet President Bush's demand for immediate withdrawal from Kuwait. Elements of the 1st and 2nd Marine Divisions, along with Army airborne troops and British, Saudi, Kuwaiti, Egyptian, and Syrian units, launched their attacks in the south, breaching Iraqi defensive lines of minefields, barbed wire, bunkers, and berms, while amphibious feint attacks under the cover of naval gunfire were launched in the Arabian Gulf. Meanwhile major Army and Coalition forces were making a "Hail Mary" end run deep into Iraqi territory before turning east to cut off Iraqi forces in Kuwait. Cut off from their supply bases and headquarters by the intense air campaign, the Iraqi occupation forces in Kuwait were already beaten. Thousands of Iraqi soldiers simply surrendered, as the Coalition forces pushed through Iraqi defenses. In a few cases some of the more elite Republican Guard units stood and fought, but their Soviet-provided equipment and tactics were no match for superior American, British, and French equipment and training.

By 26 February Coalition forces, along with the underground Kuwaiti resistance, controlled Kuwait City as Coalition air forces pounded the retreating Iraqi occupation army. In southern Iraq, Allied armored forces had reached the Euphrates River near Basra, and internal rebellions began to break out against Saddam Hussein's regime. Retreating Iraqis escaping Kuwait were on the six-lane highway (Highway 80) that runs between Iraq and Kuwait when the road was blocked at one end by Marine anti-tank mines and bombed on the other, creating a massive traffic jam of Iraqi military vehicles. On the night of 26–27 February these targets were attacked by American ground and air forces. The devastation that occurred resulted in the highway being referred to as the "Highway of Death." On 27 February President Bush ordered a cease-fire and the surviving Iraqi troops were allowed to flee into southern Iraq. On 3 March Iraq accepted the terms of the cease-fire and the fighting ended. Although Kuwait had been liberated, the U.N. sanctions against Iraq remained. The Navy would be called on to support the sanctions and limit Iraqi threats to its minorities and neighboring countries.

The No-Fly Zone

Following the Gulf War in 1991, Operation Southern Watch was carried out by the Joint Task Force Southwest Asia (JTF-SWA) to monitor and control Iraqi airspace south of the 32nd parallel. The operation began on 27 August 1992 and included forces from Saudi Arabia, the U.S., the U.K., and France as a way of enforcing U.N. resolutions to end repression of Shi'ite Muslims in southern Iraq and to prevent the Iraqi bombing and strafing attacks ordered by Hussein during the end of 1991 and into 1992. At first, the Iraqis left Coalition aircraft alone, but after the U.N. voted to maintain sanctions against Iraq, they became more belligerent.

In December an Iraqi MiG-25 Foxbat was lost to Air Force F-16s when it crossed the 32nd parallel. The following month Iraq agreed to American, British, and French demands to remove their surface-to-air missiles from below the 32nd parallel, but did not remove them all. Accordingly, President Bush ordered U.S. aircraft to strike the remaining sites. On 13 January 1993 over a hundred Coalition aircraft hit a number of SAM sites. In later months U.S. aircraft destroyed Iraqi radars that had illuminated them. Most of 1994 was relatively quiet and the Air Force began to withdraw units from the region, but in October Hussein moved two Republican Guard divisions to the Kuwaiti border after again demanding that U.N. sanctions be lifted, prompting the U.S. to rush troops to the region.[7] In June 1996 a truck bomb destroyed a barracks for American servicemen in Saudi Arabia. The Khobar Towers bombing killed 19 Americans and a Saudi and injured 372 people. Suspicion centered around Iraq, Iran, or al-Qaeda. In August 1996 Iraqi forces invaded the Kurdish areas of northern Iraq, prompting the Americans to hit targets in northern and southern Iraq under Operation Desert Strike and the no-fly zone was extended to the 33rd parallel. Again Iraqi air defenses were affected when several radars were hit by Air Force F-16s. Although France dropped out of its participation in enforcing the no-fly zones in 1998, in December of that year President Bill Clinton, citing failure to comply with U.N. Security Council resolutions, ordered Operation Desert Fox, a four-day effort against targets all over Iraq. The result was an increase in confrontations between Iraqi air defenses and Coalition aircraft. During the years following Desert Fox, there had been hundreds of incidents of missile or anti-aircraft fire at Coalition aircraft and 150 violations of the no-fly zone. By June 2002 American and British forces had stepped up attacks on Iraqi air defense targets all over southern Iraq as part of Operation Southern Focus, with the goal of degrading the Iraqi air-defense system to pave the way for a future conflict.

The Soviet Union had cooperated with the U.S. during the Gulf War, the first such joint effort since World War II. Soviet glasnost (openness) and perestroika (re-structuring) led to changes and unrest in the Soviet Union and in August 1991 a failed coup attempt triggered the dissolution of the USSR into its component states. On Christmas Day of that year Mikhail Gorbachev formally resigned as president of a Soviet Union that no longer existed. The collapse left the United States as the only world superpower, but with the end of the Cold War, regional—rather than global—threats presented themselves. The Navy's response was a new strategy that stressed littoral warfare—along the coastlines—that was set forth in a white paper ". . . From the Sea." This new global security environment led to a reduction in military forces. With the Clinton administration calling for smaller defense budgets, the Navy faced its largest drawdown since the end of World War II. But these new challenges still required the aircraft carriers to provide an American presence in the world's trouble spots.

Operation Restore Hope

In 1991 civil war had erupted in Somalia on the Horn of Africa. With the escalation of the conflict almost half of the Somali population of four and a half million were in danger of starving to death. A small force of U.N. peace-keepers was sent in to protect aid workers as part of the United Nations Operation in Somalia (UNOSOM I). With the failure of UNISOM I President Bush sent American troops to protect relief workers in an operation called Restore Hope. The *Ranger*, which had been operating in the Arabian Gulf in support of Southern Watch, was sent at high speed to the coast of Somalia on 4 December 1992 to support the massive relief effort. On 19 December she was relieved on station by the *Kitty Hawk*. With the increase in Iraqi violations of U.N. sanctions, the Hawk rushed back to the Gulf nine days later.[8] The Horn of Africa would remain an area of concern and other carriers would be sent to the area in response to crises and continuing problems with Somali pirates. In October 1993 the *Abraham Lincoln* was ordered to the coast of Somalia to assist U.N. humanitarian operations by conducting patrols over the city of Mogadishu and surrounding areas into early November in support of American ground troops during Operation Continue Hope. She was relieved on "Groundhog Station" (90 miles north of the equator in the Indian Ocean) by the *America*, which had just spent several weeks supporting U.N. peace-keeping efforts over Bosnia, only to be ordered on four hours notice to transit the Suez Canal.

The *Muavenet* Incident

During the fall of 1992 the *Saratoga* was part of Display Determination '92, a combined forces naval exercise with participants from the U.S., Turkey, and several NATO countries. Under the overall command of an American admiral, the naval forces were assigned to either of two multi-national teams. The "Brown" force, which included the *Saratoga*, was pitted against the "Green" force under a Dutch admiral. During the "enhanced tactical" phase, the Brown Force was to conduct an amphibious landing at Saros Bay in the Aegean against the opposing Green Force. The opposing forces were supposed to actively engage and "destroy" each other, in a simulated fashion, of course. On 1 October the combat direction center tactical action officer (TAO) on board *Saratoga* decided to launch a simulated attack on nearby Green Force ships using the RIM-7 Sea Sparrow. After getting approval from the *Saratoga*'s captain and the Battle Group Commander, the CDC officer initiated the simulated assault. As a "no notice" exercise, officers on *Saratoga* woke the enlisted Sea Sparrow team and told them to conduct the simulated attack. Unfortunately, not all the members of the team were aware it was only a drill. In the ensuing exercise, the language used by the TAO indicated a live firing and the missiles were "armed and tuned." Eleven minutes after midnight on 2 October, *Saratoga* fired two live missiles at the Turkish destroyer minelayer *Muavenet*, part of the Green Force only three miles away at the time.[9] The first missile hit *Muavenet* in the bridge, destroying it and the Combat Information Center. The commanding officer was among the five killed and most of the Turkish ship's officers were injured. The second missile exploded in mid-air, probably as a result of the blast from the first explosion and peppered the ship with shrapnel. Quick damage control put out the resulting fires in ten minutes and the *Muavenet* was towed to the Turkish naval base at Gölcük. Following an investigation of the incident, the captain of the *Saratoga* and seven other officers and sailors were disciplined.[10]

Bosnia

Following the breakup of Yugoslavia in 1991, the war in Bosnia began in March 1992. Several factions were involved: the forces of the Republic of Bosnia and Herzegovina and those of the Bosnian Serb and Bosnian Croat factions (Republika Srpska and Herzeg-Bosnia) who were led and supplied by Serbia and Croatia.[11] In April 1992 the United States and its NATO allies began one of the largest military operations, other than war, in the post–Cold War era. (This was also the first significant military operation ever undertaken by NATO.) These operations were in support of U.N. Security Council resolutions aimed at ending the ethnic conflicts raging in the former Yugoslavia. By the time the fighting ended in November 1995, the U.S. and its allies had flown over 109,000 sorties, only slightly less than the number flown in the Gulf War in 1991. The primary objectives of U.S. and NATO air operations over Bosnia were to enforce the maritime embargo and no-fly zone, and to deliver humanitarian aid. This involved several separate, sometimes overlapping, operations during different phases of the conflict. Although NATO aircraft were fired on by ground defenses over Serbia and Croatia, only four aircraft were lost (only one of which was American) and there was only one significant air-to-air engagement.

For much of the time at least one carrier battle group was stationed in the Adriatic Sea at what became known as "Groundhog Station."[12] At various times the task group was headed by the *America*, *Dwight D. Eisenhower*, *George Washington*, *John F. Kennedy*, *Saratoga*, and the *Theodore Roosevelt*. These carrier battle groups played a key role in enforcing the maritime embargo, providing a base for Navy and Marine Corps aircraft and coordinating U.S. and NATO air operations over the Adriatic. (Navy and Marine Corps aircraft also flew from several air bases in Italy, including Aviano, while other Marine aircraft operated from an Amphibious Ready Group in the Adriatic.)

In May 1992 the U.N. Security Council called for assistance to deliver humanitarian aid to the Bosnian population. In July, the U.S. began relief flights and by mid-July sixteen nations had landed supplies in Bosnia. In February 1993 the U.S. European Command established Joint Task Force Provide Promise under the control of U.S. Naval Forces Europe. Although the Air Force provided most of the airlift, Navy carrier-based aircraft provided air cover for cargo planes bringing in relief supplies until Operation Deny Flight was begun in April 1993. Sorties were flown by carrier-based F-14s, F/A-18s, A-6s, EA-6Bs, E-2s, and S-3s, while SH-3s provided search-and-rescue capability. In early October a U.N. Security Council resolution banned military flights in Bosnian airspace. (NATO airborne early warning aircraft that had already been involved in monitoring the maritime embargo began reporting Bosnian airspace violations.) In March 1993 the ban was extended to flights by all fixed-wing and helicopter aircraft over Bosnia and measures for compliance were authorized. NATO forces began Operation Deny Flight in April with Navy and Marine aircraft providing combat air patrol, close air support, air defense suppression, and strikes. (NATO strike missions were limited in number during this operation.) Navy E-2s provided airborne early warning and F-14s equipped with Tactical Airborne Reconnaissance Pod System (TARPS) pods were the only manned reconnaissance aircraft available to U.S. forces.[13]

In June 1993 operations supporting the embargo on weapons deliveries were combined with those enforcing the economic sanctions under Operation Sharp Guard. Carrier-based S-3s, F/A-18s, F-14s, and E-2Cs also supported these embargo operations.

In August 1995 NATO launched Operation Deliberate Force to compel the Bosnian Serbs to cease their attacks on safe areas. These air strikes, combined with successful ground campaigns by Bosnian and Croation forces and the long-term effects of the U.N. economic embargo, led to a cease-fire in Bosnia and the eventual end of the conflict through the Dayton peace process.

Beginning in December 1995 U.S. and allied nations deployed peace-keeping forces to Bosnia under Operation Joint Endeavor. (The American contribution was the 20,000 troops of Task Force Eagle.) This was the first time in NATO history that its forces were committed to a military operation as well as the first time since World War II that American and Russian soldiers shared a common mission. This was to be the last operation for the *America*, on her final deployment to the Mediterranean and Indian Ocean. One of the highlights of her last cruise was a port call at Valletta, Malta, in January 1996—the first U.S. Navy aircraft carrier to visit that historical port in over 24 years.[14]

9/11 and the War on Terrorism

On a clear Tuesday morning on 11 September 2001, 19 terrorists hijacked four passenger jets. Two of the airliners, American Airlines flight 11 and United Airlines flight 175 were deliberately flown into the north and south towers of the World Trade Center in New York. Both towers collapsed within hours; debris causing severe damage to ten other large structures in the immediate area. American Airlines flight 77 crashed into the Pentagon. The fourth airliner, American Airlines flight 93, was meant to attack the Capitol but passengers on board attempted to take control and it crashed in a field near Shanksville, Pennsylvania. Nearly 3,000 people died in the attacks, including all 227 civilians and the 19 hijackers on board the four planes. Suspicion quickly fell on al-Qaeda, an Islamic extremist group that had been advocating jihadist attacks against Western countries in response to the continued presence of American military forces in Saudi Arabia following the Gulf War, U.S. support for Israel, and the sanctions imposed against Iraq. Osama bin Laden, leader of al-Qaeda, orchestrated the attacks, although he initially denied any involvement. (After a manhunt that lasted nearly ten years, bin Laden was killed by a Navy SEAL in his compound in Abbottabad, Pakistan, on 2 May 2011.) In response, President George W. Bush announced a war on terror, aimed at bringing bin Laden and al-Qaeda to justice and preventing other terrorist networks from emerging. NATO declared that the attacks were an attack on all the member nations under the NATO charter, while Australia invoked the Australia, New Zealand, United States (ANZUS) security treaty. On 20 September President Bush delivered an ultimatum to the Taliban government of Afghanistan: turn over Osama bin Laden and al-Qaeda leaders or face attack.

Operation Enduring Freedom

At the time of the terrorist attacks, four carriers were under way. The *Enterprise* was in the Arabian Gulf and the *Carl Vinson* was in the Indian Ocean. In the Pacific the *John C. Stennis* was off the coast of California conducting carrier qualifications while the *Constellation* was on her way from Pearl Harbor to San Diego with many dependents on board for the traditional "Tiger Cruise." (The term comes from the Navy's name for dependents' cruises, Operation Tiger. They are normally at the end of a ship's deployment. It is sort of a "bring your family to work day" event since the crew's family members get to actually ride on board.) She had left the Indian Ocean in August after supporting Operation Southern Watch and, despite discussions about turning around, was allowed to complete her deployment, arriving in San Diego on 14 September. Following an abbreviated turnaround cycle, Connie prepared for her final deployment. Departing in November, she entered the Gulf in December and operated as a night carrier during the major ground combat phase. (She would also be on hand for Operation Iraqi Freedom the following March.) On 7 October 2001 the war in Afghanistan began with air strikes by the *Enterprise* and *Carl Vinson* against known al-Qaeda and Taliban camps. The carrier strikes were in coordination with Air Force B-1B Lancer, B-2 Spirit, and B-52 bombers as well as Tomahawk missiles launched by U.S. and British ships. Ground operations by Special Forces followed; by mid-November Kabul, the capital, had fallen and al-Qaeda and Taliban remnants fled back to the rugged mountains of eastern Afghanistan.[15] The *Kitty Hawk* had been ordered to leave her home port in Yokosuka, Japan, and to deploy to the North Arabian Sea in October. Having left most of her air wing in Japan, she performed a unique mission as an offshore base for U.S. Special Forces operating in Afghanistan.

More than a thousand special operations personnel, including Navy SEALs, Air Force Special Operations units, and the Army's 160th Special Operations Aviation Regiment (the "Night Stalkers") were embarked. Their specially configured equipment included a dozen MH-60 Blackhawk, six MH-47 Chinook, and several MH-53 Pave Low helicopters.[16] That same month the *Theodore Roosevelt* launched her first strikes, operating at night while the *Vinson* concentrated on daylight operations. (The *Vinson* would be relieved by the *John C. Stennis* in December.) The *Roosevelt* was relieved in March 2002 by the *Kennedy*, on her first deployment in support of Operation Enduring Freedom.[17]

Operation Iraqi Freedom

In 2002, the U.N. Security Council passed a resolution calling for Iraq to cooperate with U.N. weapon inspectors. In October of that year Congress authorized President Bush to use military force to enforce Iraqi compliance. After Saddam Hussein continued to ignore warnings to eliminate his offensive arsenal, President Bush issued an ultimatum that he and his sons leave Iraq within 24 hours. The Iraq war began on 20 March 2003 with the invasion of Iraq by the U.S. and the U.K. Before the war, the U.S. had claimed that Iraq possessed weapons of mass destruction (WMD) that posed a threat to the security of the region. Other reasons given were that Saddam Hussein was harboring and supporting al-Qaeda, that Iraq was providing financial support for Palestinian suicide bombers, that Iraqi human rights abuses were ongoing, and that the spread of democracy in Iraq must be supported. The invasion led to an occupation and the eventual capture of Hussein, who was later tried and executed by the new Iraqi government. Violence against coalition forces and among various sectarian groups led to the Iraqi insurgency, conflict between Iraqi Sunni and Shi'a groups, and the emergence of al-Qaeda in Iraq.

Five carrier battle groups, three amphibious ready groups, and two amphibious task forces (totaling over 200 coalition ships) were deployed for Operation Iraqi Freedom. The *Abraham Lincoln*, *Constellation*, and *Kitty Hawk* operated in the Arabian Sea, while the *Harry S. Truman* and the *Theodore Roosevelt* operated in the Mediterranean. On 2 April the *Kitty Hawk* suffered a friendly fire incident when an F/A-18C flown by Lieutenant Nathan D. "O.J." White of strike fighter squadron VFA-195 was killed by an Army Patriot missile battery, which mistook him and his wingman for hostile Iraqi aircraft while they were flying a night close air support mission over Karbala.[18] In July the *Kennedy*, after participating in a major inter-service exercise, transited the Suez Canal and entered the Red Sea to come under Fifth Fleet operational control. She provided cover for support for the multi-national forces in Iraq through December. It would be her last deployment.[19]

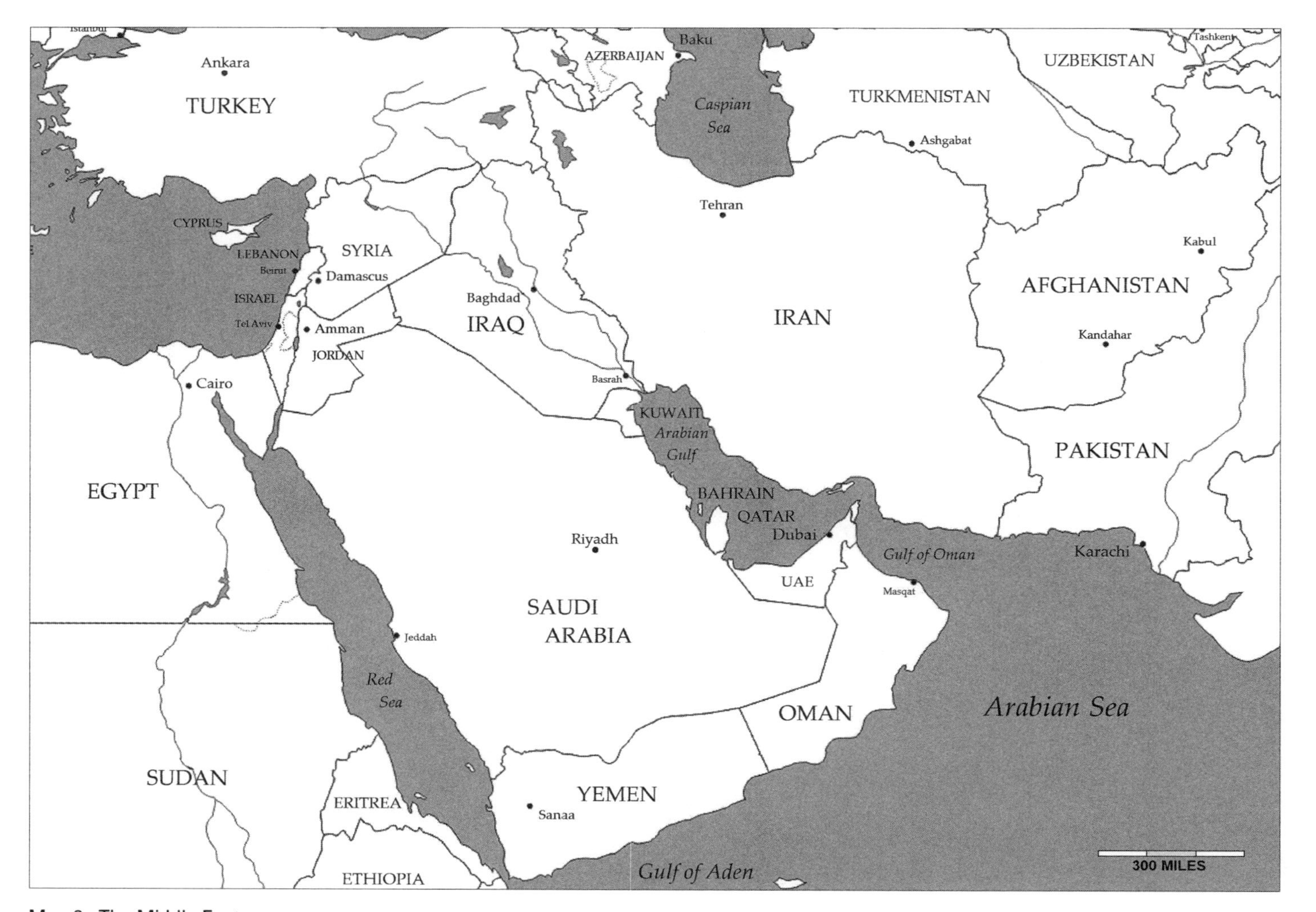

Map 6. The Middle East

Map adapted by the author from (South-West Asia) http://d-maps.com/carte.php?num_car=29000&lang=en.

14 Into the Sunset

In 1966 Defense Secretary Robert McNamara, having been convinced of the usefulness of nuclear power for warships, included a new nuclear carrier in the FY67 defense budget. In many ways the design owed as much to the experience gained with the *Forrestal* and *Kitty Hawk* classes as to her nuclear-powered predecessor, the *Enterprise*, which was destined to remain, quite literally, in a class by herself. The *Nimitz* was commissioned in 1975 and became the pattern for the ships that followed: the *Eisenhower* was commissioned in 1977, the *Carl Vinson* in 1982, and the *Theodore Roosevelt* in 1986. As part of the Reagan buildup under Navy Secretary John Lehman, these ships were followed by the *Abraham Lincoln* in 1989, the *George Washington* in 1992, the *John C. Stennis* in 1995, and the *Harry S. Truman* in 1998. Later ships of the *Nimitz* class included the *Ronald Reagan* commissioned in 2003 and the *George H.W. Bush* in 2009.[1] As these new ships entered the fleet, the older conventionally powered carriers reached the end of their useful lives and were decommissioned. Their situation was exacerbated by an unrelenting operational tempo and the tendency to cut funding for scheduled maintenance and overhauls. In the leadership climate of the times, there was also the unfortunate tendency of senior leaders to "gundeck" routine inspections, further accelerating the decline in their material condition. (In modern Navy usage, gundecking is the falsifying of reports, records, and the like to avoid failing inspections while not actually having to follow proper procedures or perform required actions.)

The Naval Vessel Register (NVR) is the official inventory of ships in the U.S. Navy. Vessels are listed in the NVR when they are authorized to be built and remain in the NVR throughout their life as Navy assets. Ships that have been decommissioned may become part of the "reserve fleet" for retention as mobilization assets until needed by the active fleet. These Reserve Category B ships are designated by the Chief of Naval Operations to receive the highest degree of maintenance practicable within personnel and funding limitations. Such "mothballing" includes dehumidification equipment, cathodic protection (to guard against corrosion), and flooding alarms. The NVR also lists ships that have been stricken but not disposed of as well as their final disposition. (The term "stricken" means that a ship has been formally removed from the register by the Secretary of the Navy on the recommendation of the CNO; it is a legal preliminary to disposal.) Reserve Category X applies to ships that have been stricken from the NVR and are awaiting disposal by scrapping, sale to foreign countries, or designation as targets, memorials, or donations. They receive no maintenance or preservation, only security against fire, flooding, and pilferage.[2] The formal term for scrapping a ship is "dismantling." In order to ensure that they are properly demilitarized, the Navy has always dismantled its warships in the United States and that is not likely to change. In 1999 the Navy started the Ship Disposal Project (SDP) to get a handle on the scope and associated costs of ship dismantling and recycling. Under the program, contractors tow an inactive ship to their facility and do all the work of removal and proper disposal of hazardous materials, and the recycling of scrap metals and salvageable equipment. This cost has always been borne by the Navy, but in the future, contracts will be executed when scrap metal commodity prices in the market allow dismantling to be more cost effective. Because of these considerations, inactive ships might linger in a kind of official limbo for years while awaiting their ultimate fate. For a few fortunate ships, scrapping is not the end. Since 1948 the Navy's Ship Donation Program has made selected historically significant naval vessels available for donation to non-profit organizations, states, or cities. This policy restricts the use of donated naval vessels for the purpose of public display as a museum or memorial. When transfer occurs, the recipient assumes title to the vessel and all ownership costs of restoration, preservation, maintenance, operation as a static museum or memorial for public display, and, ultimately, for final disposal. Because of these financial and organizational hurdles, it is a difficult and drawn out process for any organization to be ultimately successful in such preservation efforts.

Ranger

The *Ranger* was the first to go. In April 1992 she had participated in a re-enactment of the Doolittle Raid on Tokyo, Japan, as part of the World War II 50th anniversary commemorative activities. Two World War II–era B-25 bombers were craned on board and, as over 1,500 embarked guests watched, the two vintage warbirds thundered down her flight deck and took off. Following a port visit to Vancouver, British Columbia, that June, she began working up for her 21st and final deployment to the Pacific and the Indian Ocean. Departing in August, she stopped at Yokosuka, Japan, for a week of upkeep before heading for the Indian Ocean in September. Transiting the Straits of Hormuz to enter the Arabian Gulf, she relieved the *Independence* in an unusual close aboard ceremony and immediately began flying missions in support of Operation Southern Watch. While in the Gulf, the *Ranger*, along with British and French naval forces, joined the Russian guided-missile destroyer *Admiral Vinogradov* in communication, maneuvering, and signaling drills. During the exercise a Russian Ka-27 Helix helicopter landed on board, the first such landing on a U.S. Navy carrier. In December she left the Gulf and steamed at high speed to the coast of Somalia, where she supported the massive relief effort for starving Somalis as part of Operation Restore Hope. Later that month she was relieved on station by the *Kitty Hawk* and returned to San Diego at the end of January 1993. She was to have undergone the Service Life Extension Program (SLEP), which would have extended her service life to 2002, but the Navy changed its plans and she was officially decommissioned on 10 July 1993 after 46 years of service. The *Ranger* was transferred to the Naval Inactive Ship Maintenance Facility, Bremerton, Washington. Her "parking spot" is available until 2014, after which she will be sold for scrap.[3] The not-for-profit USS Ranger Foundation submitted an application to Naval Sea Systems Command (NAVSEA) proposing the donation of *Ranger* for use as a museum ship and multi-purpose facility on the Columbia River at Chinook Landing Marine Park in Fairview, Oregon.[4] In September 2012 NAVSEA canceled her donation hold, but the Northwest Historic Ships Association is trying to move her to Everett, Washington, before then.[5]

Forrestal

During the first months of 1991 the *Forrestal* was working up as the East Coast readiness carrier in anticipation of deploying to support Operation Desert Storm, but these plans were canceled twice. She finally departed on her 21st and final deployment at the end of May. During the next seven months of this "transitional" deployment, *Forrestal* supported Operation Provide Comfort and participated in testing and evaluating new carrier roles and Sixth Fleet battle group tactics. It was during her support of Provide Comfort that a bizarre incident occurred. On 8 July an E-2C from VAW-122 developed an engine fire that soon burned out of control. Although the five-man crew had bailed out safely, the aircraft was still headed toward Syrian airspace. Lieutenant William Reilly of VFA-132 was ordered to destroy the E-2, which he shot down with his Hornet's 20mm cannon. Ironically, this would be the *Forrestal*'s first and only aerial "kill" in her entire history. At the end of her final deployment in December, the *Forrestal* was selected to replace the *Lexington* as the Navy's training carrier and was redesignated AVT-59. Arriving at her new home port of Pensacola, Florida, in February 1992, she conducted limited training operations until she was ordered to the Philadelphia Navy Yard in September for an extensive overhaul. The Navy had intended that she could be returned to service as an operational carrier within a year if needed, but she too fell victim to budget cuts and the overhaul was canceled in mid-1993 when the Navy decided to do without a dedicated training carrier. She was decommissioned on 11 September 1993 and stricken from the Naval Vessel Register the same day, after which she was stripped to support her sisters still on active service. In 1999 a campaign was started by the USS Forrestal Museum Inc. to obtain the ship for use as a museum in Baltimore, Maryland, but this plan fell through and NAVSEA canceled her donation hold in 2004 and put her up for disposal. In June 2010 *Forrestal* was towed from the Naval Station in Newport,

Rhode Island, where she had been stored since 1998, for the inactive ship storage facility in Philadelphia. She is tied up next to the *John F. Kennedy* and will remain there until disposed of by dismantling.[6]

Saratoga

Sara began her final deployment to the Mediterranean in January 1994, turning over from the *America* east of Gibraltar before entering the Adriatic on 1 February where she launched the first of thousands of sorties in support of the U.N. and NATO operations Deny Flight and Provide Promise over Bosnia-Herzegovina. After a port call to Trieste, Italy, *Saratoga* departed at the end of the month and resumed station in the Adriatic until 10 March before heading for the eastern Mediterranean for land and naval exercises. She returned to Trieste, then to the Greek island of Crete for bombing exercises at the Avgo-Nisi bombing range. Completing these exercises, she returned to the Adriatic for five more days of flying in support of Deny Flight and Provide Promise, departing Groundhog Station 7 April; she transited the Straits of Messina between Sicily and the tip of southern Italy, for a port call at Naples before returning to the Adriatic for the fourth time on 17 April. While operating in the Adriatic on 28 April one of her F/A-18 Hornets from the VA-83 Rampagers crashed in the Adriatic during takeoff, killing the pilot. This was the first death among the NATO allies supporting Operation Deny Flight. In early May *Saratoga* anchored off Palma de Mallorca for a week of liberty at the resort city before getting under way again to participate in the exercise Dynamic Impact involving joint maneuvers with several NATO navies and the U.S. Air Force. After another liberty call in Valencia, Spain, she participated in the Iles D'Or (Islands of Gold) exercise with the French Navy until 9 June. The *Saratoga* rendezvoused with her relief, the *George Washington* before returning to Mayport, where she arrived on 24 June 1994.

On 20 August 1994 *Saratoga* was decommissioned at the Naval Station, Mayport, and stricken the same day. In May 1995 she was towed to Philadelphia. When the Philadelphia Navy Yard was deactivated in August 1998, she was taken to the Newport Naval Station where she was placed on donation hold. Her status was changed to "disposal as an experimental ship" but she was returned to donation hold status on 1 January 2000. Like her sisters, she has been extensively stripped to support the active carrier fleet. There was an effort to make her a museum ship in Quonset Point in North Kingstown, Rhode Island, but in April 2010 she was again removed from donation hold and scheduled for disposal. In 1994 and 1995 efforts to make her a museum ship in Jacksonville, Florida, failed to raise even half of the startup costs. Although Jacksonville civic leaders tried to raise funds with a "Save Our Sara" campaign, it fell short of the $3 million goal. When startup costs increased from $4.5 million to $6.8 million, efforts were abandoned. (A major difficulty for Jacksonville was competition with the National Football League, who had awarded the city a franchise for the Jacksonville Jaguars in November 1993. The city had to ensure a large financial commitment to fund re-building of the city's stadium as part of the deal, severely limiting the city's available funding and support.) The

Jacksonville USS Saratoga Museum Foundation, Inc. stopped operating in the summer of 1995. The *Saratoga* is still in Newport while the Navy is reviewing bids to scrap her.[7]

America

In August 1995 *America* left Norfolk for her 20th and final deployment to the Mediterranean, the Adriatic, and the Persian Gulf. She supported U.N. and NATO operations Deny Flight and Operation Deliberate Force, and also flew missions in support of Operation Southern Watch over Iraq. She supported the NATO Implementation Force in Bosnia and Herzegovina for Operation Joint Endeavor before returning to Norfolk in February 1996.

America was scheduled to undergo a Service Life Extension Program (SLEP) in the late 1990s, but fell victim to budget cuts and was retired early since she was in such poor material condition. In the early 1990s one of her flight deck elevators fell with an S-3 Viking and several plane handlers on it. She suffered steam and fuel leaks and—while returning home from a deployment—plowed through a hurricane that destroyed large sections of the flight deck catwalks.[8] On 9 August 1996 she was decommissioned in a ceremony at Norfolk Naval Shipyard in Portsmouth, Virginia, after which she was stricken and transferred to the Ready Reserve Fleet at the Inactive Ship Maintenance Facility in Philadelphia. Though already decommissioned, she was awarded the 1995 Battenberg Cup in recognition of her crew's achievements in her last full year in service.

It was planned that *America* would be sold for scrap, but she was instead selected for a live-fire test and evaluation in 2005. Some objected to a ship named for the nation being deliberately sunk at sea, and a committee of former crew members and other supporters attempted to save her as a museum. Their efforts were ultimately unsuccessful. The Navy's response was that *America* could "make one final and vital contribution" to national defense, in providing data for the design of future carriers.[9] On 25 February 2005 a ceremony to salute *America* and her crew was held in Philadelphia, attended by former crew members and various dignitaries. She departed the Inactive Ship Maintenance Facility on 19 April 2005. The tests, which lasted about four weeks, included explosions, both underwater and above the surface, designed to simulate torpedo, cruise missile, and small boat suicide attacks. Monitoring devices were placed on the ship to allow results to be transmitted to the vessels recording the data. When the tests were complete, *America* was sunk in a controlled scuttling on 14 May 2005 around 250 miles southeast of Cape Hatteras. The sinking was not publicized until six days later. A Freedom of Information Act request from former *America* veterans resulted in the Navy releasing the exact location. She now lies upright in one piece, 16,860 feet below the surface at 33°09′09″N 71°39′07″W.[10]

Independence

In September 1991 the *Independence* changed her home port to Yokosuka, Japan, when she replaced the *Midway* as the Navy's only permanently forward-deployed aircraft carrier.[11] In her first deployment to the Indian Ocean and the Persian Gulf following her change in home ports, she became the first carrier to enforce the southern no-fly zone in August 1992 when Operation Southern Watch began. Over the next few years the Indy would make four deployments to the Gulf in support of Southern Watch.

In June 1995 the *Independence* became the oldest ship in the active fleet and, as such, was authorized to fly the First Navy Jack, known as the "Don't Tread on Me" flag. (The jack is flown from the bow of a ship when in port. The standard Navy jack was the blue field with white stars of the U.S. flag. After 11 September 2001 the Secretary of the Navy directed that all U.S. Navy ships raise the historic jack and continue to do so throughout the global war on terrorism.)

In mid- to late 1995 the People's Republic of China (PRC) conducted a series of missile tests to intimidate the Taiwanese government from moving away from the "One China" policy. A second set of missiles were fired early in 1996 in an apparent attempt to influence the upcoming Taiwanese presidential elections. President Clinton responded by sending the *Independence* to the waters east of Taiwan in March 1996 to provide a stabilizing presence during what became known as the Third Taiwan Strait Crisis. She was joined in the area by the *Nimitz* as the PRC continued to fire missiles into Taiwanese territorial waters. Upon her return to Yokosuka in April, the ship was visited by President Clinton as part of an official state visit to Japan.

During a four-month deployment in 1997 *Independence* participated in several major exercises and made seven ports of call, which included one to Guam in February (the first aircraft carrier to pull into Guam in 36 years) and another in April to Port Klang, Malaysia (becoming the first aircraft carrier of any navy to visit Malaysia). Before returning to Yokosuka, she made her last port call in May 1997 to Hong Kong, the last U.S. naval port visit to the territory before its handover to China in July 1997. In January 1998 *Independence* made her last deployment to the Gulf to support Southern Watch before departing for the West Coast in June.

On 30 September 1998 she was decommissioned in a ceremony at Puget Sound Naval Shipyard in Bremerton, Washington. After her decommissioning her "Don't Tread on Me" First Navy Jack was transferred to the *Kitty Hawk*. She was in moth balls for four and a half years before being struck in March 2004. Like her sisters before her, she was heavily stripped to support the active carrier fleet. Her port anchor and both anchor chains were used on the *George H.W. Bush*. In April 2004 she was named as one of 24 decommissioned ships available to be sunk as artificial reefs. Until a contract is awarded for her scrapping, she remains available for donation as a reef.

Constellation

The *Constellation* completed her three-year Service Life Extension Program (SLEP) at Philadelphia in March 1993. After completing her shakedown exercises, she departed Mayport in late May and, while transiting to the Pacific, operated with various Latin American air forces along the way. She arrived at her home port in San Diego in mid-July. After participating in Rim of the Pacific (RIMPAC) exercises from May to June 1994, she left San Diego in November for her first extended deployment in six years. She spent most of December in WestPac participating in exercises off Okinawa, followed by a number of exercises off Korea when world attention was focused on the news that North Korea was attempting to develop nuclear weapons. In January 1995 she was in the Persian Gulf patrolling the no-fly zone for Operation Southern Watch. Her six month deployment ended in May with her return to San Diego. She returned to the Gulf again to support Southern Watch during her April to October 1997 deployment. While preparing for her 1999 deployment, tensions in Korea were again on the rise following an exchange of gunfire between North and South Korean naval vessels. Connie left San Diego in mid-June for the Korean Peninsula to monitor the situation. In August she entered the Persian Gulf and again supported Southern Watch, during which she conducted air strikes against two Iraqi radar stations. In September an F-14 from the VF-2 Bounty Hunters attempted to engage an Iraqi jet with a long-range Phoenix air-to-air missile. In all, Connie's CVW-2 aircraft engaged in nine ordnance-dropping air strikes while in the Gulf. Departing in November she arrived home in time for the holidays in December. In March 2001 *Constellation* left for yet another deployment to the Persian Gulf and immediately after entering the Gulf at the end of April began flight operations in support of Southern Watch, during which she conducted air strikes in response to Iraqi violations of the no-fly zone. Connie left the Gulf in early August for Pearl Harbor, Hawaii, departing on 9 September with dependents on board for the traditional Tiger Cruise on the final leg to San Diego. The *Constellation* was almost halfway to San Diego when word of the terrorist attacks on New York and the Pentagon was received. The carrier was allowed to complete her return and arrived in San Diego on 14 September. The following month *Constellation* observed her fortieth anniversary and in early November 2002 departed for her final deployment to the Western Pacific, the Indian Ocean, and the Arabian Gulf supporting the operations Enduring Freedom and Iraqi Freedom. She was designated a night carrier and remained on station throughout the major ground combat phase, launching more than 1,500 sorties and her aircraft delivered over 1.7 million pounds of ordnance. Although one aircraft was lost in an operational accident, there were no fatalities.

Connie left the Gulf in April 2003, returning to San Diego for the last time early in June. She was decommissioned at the North Island Naval Air Station on 7 August 2003. In September she was towed to the Naval Inactive Ship Maintenance Facility, Bremerton, Washington, and stricken on 2 December 2003. Currently, *Constellation* is scheduled to be disposed of by dismantling.

John F. Kennedy

After a brief period of maintenance in Mayport in early 2000, the *Kennedy* sailed north to New York to participate in the International Naval Review over the 4th of July holiday, and then headed to Boston for Sail Boston 2000. After returning to Mayport she was refitted as a test bed for the Cooperative Engagement Capability, which allowed the ships and aircraft of her battle group to share sensor data and provide a single, integrated operational picture. As the 9/11 terrorist attacks unfolded, *Kennedy* and her battle group were ordered to support Operation Noble Eagle, establishing air security along the mid-Atlantic seaboard, including Washington, D.C. JFK was released from Noble Eagle on 14 September 2001. Her next scheduled deployment was accelerated by three weeks in response to the terrorist attacks and she departed in early February 2002 to operate in the Arabian Sea. During the first six months of 2002 *Kennedy* aircraft bombed Taliban and al-Qaeda targets in support of Operation Enduring Freedom. Upon returning in August, Big John served as the east coast ship for carrier qualifications. She began an extensive maintenance period from January to October 2003, the largest pierside maintenance ever conducted outside of a major naval shipyard. *Kennedy* got under way in June 2004 in support of operations Iraqi Freedom and Enduring Freedom. In July 2004 *Kennedy* collided with a dhow in the Arabian Gulf, leaving no survivors. (After the incident the Navy relieved her commanding officer.) The carrier itself was undamaged, but two fighters on the flight deck were damaged when one slid into the other as the ship made a hard turn to avoid the Arab vessel.

As the most costly carrier in the fleet to maintain, budget cutbacks and changing naval tactics prompted the Navy to cancel her 15-month extensive overhaul in April 2005 and it was announced that the Navy would retire her. As a last farewell, *Kennedy* made a number of port calls, including a stop at her home port in Boston as well as participating in the New York City Fleet Week festivities at the Intrepid Sea-Air-Space Museum for one last time. A decommissioning ceremony was conducted in Mayport, Florida, on 23 March 2007 and she was towed to Norfolk, Virginia, in July. She remained there until preparations could be made to tie her up at the Naval Inactive Ship Maintenance Facility in Philadelphia. In November 2009 the *Kennedy* was placed on donation hold for use as a museum and memorial; a newspaper report in the *Boston Herald* mentioned the possibility of bringing her to Boston at no cost to the city. In August 2010 two groups—the Rhode Island Aviation Hall of Fame in Providence, Rhode Island, and the USS John F. Kennedy Museum in Portland, Maine—had successfully completed the initial phase of the Navy's Ship Donation Program, but in January 2011 the Portland City Council voted to not continue with the project. In the meantime, the ship's unique in-port cabin, which had been decorated by Jacqueline Kennedy, was disassembled and sent to the National Museum of Naval Aviation at Naval Air Station Pensacola, Florida, where it is currently in storage. With its wood paneling, oil paintings, and rare artifacts, it is a unique historical asset that may be restored on board if she becomes a museum.[12]

Kitty Hawk

In July 1998 the *Kitty Hawk* left San Diego to take over from the *Independence* as the only permanently forward-deployed carrier, arriving in Yokosuka in August. In September 1998, with the decommissioning of the *Independence*, *Kitty Hawk* became the oldest active ship (after the sail frigate *Constitution*) and was authorized to fly the First Navy Jack. Adopting Air Wing 5, she departed for the Arabian Gulf the following March, racking up over a thousand combat sorties in support of the no-fly zone, returning in August 1999. In 2000 the Hawk conducted local operations, but supported Operation Enduring Freedom in 2001.[13]

In 2002 the Hawk completed two three month deployments in the Western Pacific, but in January 2003, as she began another WestPac deployment, she would again be called to the Indian Ocean in February to support Operations Southern Watch and Iraqi Freedom. Over the next few years she would make several deployments, highlighted by port calls to Hong Kong, South Korea, Singapore, and Australia, with occasional periods of restricted availability for upkeep. In late October 2006, as *Kitty Hawk* and her escorts were conducting exercises near Okinawa, a Chinese *Song*-class submarine shadowed the group and then surfaced within five miles of the battle group. Some reports claimed that the submarine had been undetected until it surfaced, raising some unsettling questions about Chinese intentions and capabilities. The Hawk continued to operate in the Western Pacific, but in May 2008 she left Japan to turn over to the *George Washington*, her replacement as the forward-deployed carrier. But a fire on the *George Washington* caused a change in plans and the *Kitty Hawk* took her place for the RIMPAC exercises off Hawaii. Once the turnover was complete, she headed for San Diego, arriving in early August. She departed San Diego with a reduced crew and a contingent of former crew members (including some who were "plank owners") for Bremerton, where she arrived in early September. *Kitty Hawk* was formally decommissioned on 12 May 2009. The Navy will maintain her in reserve until 2015, when the *Gerald R. Ford* is commissioned. In the meantime, a group in Wilmington, North Carolina, is working to bring the ship to the city after that to serve as a museum alongside the battleship *North Carolina*.

Epitaph

The *Forrestal* and her sisters have served America for over half a century and their careers have spanned multiple wars and crises around the world. Their very existence created the concept of the "super carrier" that continues through their nuclear-powered successors. They proved themselves capable of adapting to new technology, new doctrine, and new requirements. Although the need for such ships as aircraft carriers has been debated many times over the years, whenever there was a critical situation, they stood ready to answer the question: "Where are the carriers?"

Acknowledgments

I wish to express my appreciation to the following individuals and organizations, who have given generously of their time and advice, for their assistance and encouragement in writing this book. Dale J. "Joe" Gordon and the staff at the Naval Historical and Heritage Command were instrumental in granting me access to the voluminous information available at the Washington Navy Yard. Others who made significant contributions include fellow author Norman Polmar, Navy Memorial archivist Robert C. Smith, Mike Dillard from Newport News Shipbuilding, and Tailhook Association historian Doug Siegfried. A special thanks also to the staff at Nautical & Aviation and to my editor, Captain Rosario M. "Zip" Rausa, USN (Ret.), as well as Susan Brook and Emily Bakely of the Naval Institute Press. I would also like to thank my wife Sherryl and my sons James, Christopher, and David for putting up with me while I struggled through this book. Finally, I would like to acknowledge a former co-worker, Lieutenant Colonel John Jamka, USA, whose repeated inquiry, "So how's the book coming?" prompted me to stick with it until completion.

Appendix A

Technical Data

Designing a ship is a series of compromises; it must be strong enough structurally to withstand the forces of the sea that act upon it while having inherent qualities of buoyancy and stability. In warship design, competing requirements for armament, protection, sea worthiness, maneuverability, speed, endurance and habitability must be balanced. For aircraft carriers, their size is ultimately dictated by the numbers and types of aircraft they will operate. The *Forrestal* and her sisters had the growth potential to adapt to new roles and missions. The basic soundness of their design can be seen in the fact that the nuclear-powered carriers that succeeded them are not appreciably different in general layout or in their overall dimensions.

Dimensions:

Carrier	Std. Displ. (tons)	Full load Displ. (tons)	Std. Draft (ft-in)	Length Overall (ft-in)	Length Waterline (ft-in)	Beam Waterline (ft-in)	Width Overall (ft-in)
CV-59 *Forrestal*	59,650	78,000	37'0"	1,039'0"	990'0"	126'4"	252'0"
CV-60 *Saratoga*	60,000	78,000	37'0"	1,039'0"	990'0"	126'4"	252'0"
CV-61 *Ranger*	60,000	78,000	37'0"	1,039'0"	990'0"	126'4"	259'10"
CV-62 *Independence*	60,000	78,000	37'0"	1,046'0"	990'0"	126'4"	252'0"
CV-63 *Kitty Hawk*	60,100	80,000	35'9"	1,062'8"	990'0"	126'4"	249'4"
CV-64 *Constellation*	60,100	80,000	35'9"	1,072'6"	990'0"	126'4"	249'4"
CV-66 *America*	60,300	80,000	35'9"	1,047'7"	990'0"	129'11"	249'4"
CV-67 *John F. Kennedy*	61,100	80,000	35'9"	1,047'7"	990'0"	129'11"	252'3"

Propulsion:

Propulsion for the *Forrestal*, as first ship in her class, was provided by a 600 psi (pounds per square inch) 260,000 shp (shaft horsepower) steam-turbine plant with eight Babcock & Wilcox boilers and four steam turbines driving four 22-foot diameter five-bladed propellers. Steering was provided by twin rudders. (As much of the original design was inherited from the canceled *United States*, both the *Forrestal* and *Saratoga* had a third centerline rudder as an extension of the keel. Whether these ships ever operated with the third rudder is unclear from the sources available, but at some point, they were welded in place.) All subsequent ships had 1,200 psi systems that provided 280,000 shp. (The 1,200 psi boiler systems were introduced in 1954 and offered higher efficiency, reduced weight, smaller volume and simplified maintenance over the 600 psi systems of World War II vintage.) Propulsion for the *Kitty Hawk* class was provided by four Westinghouse steam turbines and eight Foster Wheeler boilers. As designed all these ships were capable of 33 knots. In the 1970s all Navy surface ships using Navy Special Fuel Oil (NSFO) were converted to use Naval Distillate Fuel (NDF).

Arresting Gear:

The *Forrestal*-class carriers were fitted with Mark 7 systems capable of stopping a 50,000-pound aircraft (up to 60,000 pounds in an emergency) at 105 knots (121 mph). Originally there were six cross-deck pendants, but this was later reduced to four in the early 1960s. The last 120 feet on the angled flight deck on the early ships was widened by 15 feet when improved arresting gear was installed to accommodate the longer run out. The *Kitty Hawk* and *Constellation* were fitted with five Mark 7 Mod. 2 systems. (The fifth system is used for the barricade, which is rigged when needed.) The *America* and *John F. Kennedy* were fitted with Mark 7 Mod. 3 systems. The *Forrestal*, *Saratoga*, and *Independence* received Mod. 3 systems during their Service Life Extension Program (SLEP) modernizations.

Catapults:

The *Forrestal* and *Saratoga* were built with two C-7 steam catapults on the bow forward and two C-11 catapults on the port angled deck sponson. Later ships of the *Forrestal* class, the *Ranger* and *Independence* were equipped with four C-7 catapults. The *Kitty Hawk* and *Constellation* were built with four improved C-13 catapults, while the *America* and *John F. Kennedy* had three C-13 and one C-13 Mod. 1 (which had a longer track length; see table). Other upgrades to the aircraft launching systems include improved Jet Blast Deflector (JBD) equipment, particularly for the F-14 Tomcat.

Catapult	C-7	C-11/C-11-1	C-13	C-13-1
Stroke	253′	211′	249′10″	309′8.75″
Track Length	276′	225′	264′10″	324′10″
Tractor and Piston Weight	5,200 lbs	5,200 lbs	6,350 lbs	6,350 lbs
Cylinder Diameter	18″	18″	18″	18″
Stroke Volume (cu ft)	944	786	910	1,148

Carrier	Catapults
CV-59 *Forrestal*	Two C-7; two C-11
CV-60 *Saratoga*	Two C-7; two C-11
CV-61 *Ranger*	Four C-7
CV-62 *Independence*	Four C-7
CV-63 *Kitty Hawk*	Four C-13
CV-64 *Constellation*	Four C-13
CV-66 *America*	Three C-13; one C-13 Mod 1
CV-67 *John F. Kennedy*	Three C-13; one C-13 Mod 1

Armament:

5"/54 The Mark 42 5"/54 caliber automatic, dual-purpose (air/surface target) gun mount, normally controlled remotely from a Mark 68 Gun Fire Control System or locally from the mount at the One Man Control (OMC) station, had a maximum rate of fire of 40 rounds per minute; the maximum range was about 13 nautical miles and the maximum altitude was about 50,000 feet.

Terrier The RIM-2 Terrier surface-to-air missile (SAM) was a two-stage, medium-range missile semi-active radar homing system with a range of 40 nautical miles at speeds as high as Mach 3. It was launched from a Mark 10 Mod. 3 or 4 launcher. (The Mod. 3 was on the starboard quarter and the Mod. 4 on the port quarter.) The associated SPG-55 guidance radars were removed when the Terriers were replaced.

Sea Sparrow The AIM-7E Sparrow was adapted to become the Point Defense Missile System using existing antisubmarine rocket (ASROC) launchers. The radar was a manually trainable unit with continuous wave illuminators. The *Forrestal* Sea Sparrow launcher was added to a sponson on the starboard side forward in 1967. Two launchers were fitted to the *Independence* in 1973, two on the *Saratoga* in 1974, and two on the *Forrestal* in 1976. The *Ranger* was fitted with a system but retained her after two 5" guns until 1977. In cooperation with NATO, the Sea Sparrow has been

Carrier	Year	Armament	Notes
CV-59 *Forrestal*	1955	8 Mk 42 5"/54 gun mounts	
	1962	4 Mk 42 5"/54 gun mounts	
	1968	1 Mk 25 BPDMS Sea Sparrow launcher	
	1976	2 Mk 25 BPDMS Sea Sparrow launchers	
	1985	2 Mk 29 IPDMS NATO Sea Sparrow launchers	
	1985	3 Mk 15 CIWS gun mounts	
CV-60 *Saratoga*	1956	8 Mk 42 5"/54 gun mounts	
	1963	4 Mk 42 5"/54 gun mounts	
	1975	2 Mk 25 BPDMS Sea Sparrow launchers	
	1980	3 Mk 15 CIWS gun mounts	
CV-61 *Ranger*	1957	8 Mk 42 5"/54 gun mounts	
	1966	4 Mk42 5"/54 gun mounts	
	1975	2 Mk 42 5"/54 gun mounts	Sponsons retained
		3 Mk 25 BPDMS Sea Sparrow launchers	
	1978	2 Mk 29 IPDMS NATO Sea Sparrow launchers	
		3 Mk 15 CIWS gun mounts	
CV-62 *Independence*	1959	8 Mk 42 5"/54 gun mounts	
	1961	4 Mk 42 5"/54 gun mounts	
	1978	2 Mk 29 IPDMS NATO Sea Sparrow launchers	
		3 Sea Sparrow	
		3 Mk 15 CIWS gun mounts	
CV-63 *Kitty Hawk*	1961	2 Mk 10 Mod 3/4 RIM-2 Terrier launchers	
	1978	2 Mk 29 IPDMS NATO Sea Sparrow launchers	
		2 Mk 15 CIWS gun mounts	
		2 Mk 49 RAM	CIWS removed
CV-64 *Constellation*	1961	2 Mk 10 Mod 3/4 RIM-2 Terrier launchers	
	1978	2 Mk 29 IPDMS NATO Sea Sparrow launchers	
		3 Sea Sparrow	
		3 Mk 15 CIWS gun mounts	
CV-66 *America*	1965	2 Mk 10 Mod 3/4 RIM-2 Terrier launchers	
	1980	3 Mk 29 IPDMS NATO Sea Sparrow launchers	
	1980	3 Mk 15 CIWS gun mounts	
CV-67 *John F. Kennedy*	1968	3 Mk 25 BPDMS Sea Sparrow launchers	
	1981	3 Mk 29 IPDMS NATO Sea Sparrow launchers	
		2 Mk 15 CIWS gun mounts	
		2 Mk 49 RAM	CIWS removed

continually updated through the Improved Point Defense Missile System (IPDMS) and the Evolved Sea Sparrow missile (ESSM). *Saratoga*, *Forrestal*, *Independence*, *Kitty Hawk*, and *Constellation* received the improved NATO Sea Sparrow systems during their SLEP overhauls.

Close-In Weapons System The Phalanx Close-In Weapons System (CIWS) (usually pronounced "Sea Whiz") is an anti-ship missile defense system with a radar-guided six-barrel 20mm gatling gun mounted on a swiveling base. It was developed from the M61 Vulcan aircraft weapon and was first operational on board the *America* in 1980. With a rate of fire of 3,000 to 4,500 rounds per minute and capable of fast elevation and traverse speeds to track incoming targets, CIWS is an entirely self-contained unit. The gun, automated fire control system, and all other major components are included on the mount, enabling it to search, detect, track, and engage targets automatically. As both the threat and computer technology have evolved, CIWS has been upgraded to improve its capabilities against surface targets as well as supersonic anti-ship missiles and now includes infrared sensors and links to the Rolling Airframe Missile (RAM) system.

Rolling Airframe Missile The RIM-116 Rolling Airframe Missile (RAM) is a small, lightweight, infrared homing surface-to-air missile originally intended as a point-defense weapon against anti-ship cruise missiles. (A RAM missile rolls around its longitudinal axis to stabilize its flight path, not unlike a rifle bullet.) The RAM system was first operational in 1992 and is integrated with the ship's defense systems. Only the *Kitty Hawk* and *Kennedy* were each equipped with two RAM systems, which replaced their CIWS systems.

Radar:

Aircraft carriers have a myriad of electronic systems that perform a multitude of functions. These include radars, communications transmitters and receivers, navigation systems, fire control systems, electronic warfare equipment, and various countermeasures. These systems and their components are identified by a common nomenclature system, the Joint Electronics Type Designation System (JETDS). In JETDS, equipment sets or systems are prefixed by the letters "AN/" and followed by three letters then a hyphen, a number, and sometimes some other letters. The first letter indicates the platform the equipment is installed on, the second the type of equipment, and the third the purpose. The letters are followed by a model number. For example, the SPS-10 is a shipboard surface search radar. (See illustration.) The numbers are assigned sequentially, so higher numbers indicate more modern systems. Letters are used to show different versions of the same equipment, e.g., SPS-43A. (JETDS is described by MIL-STD-196.)

Shipboard radars generally fall into two categories: surface search and air search. (There are also specialized radars for controlling aircraft or weapons. These are covered under their associated systems in the text.) Air search radars during World War II were two dimensional (2D), that is, scanning in range and bearing, while separate

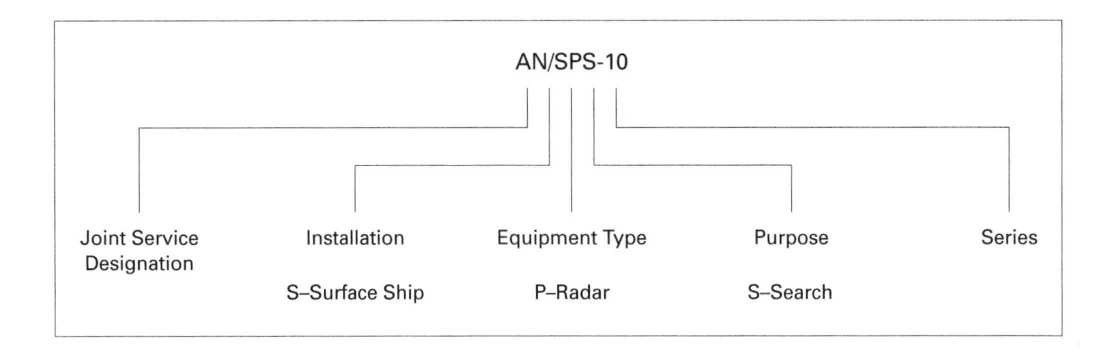

height finding (HF) radars were used to determine the elevation, and thus the altitude, of airborne targets. In the 1960s, the Navy introduced three dimensional (3D) radars that scanned in range, bearing, and elevation. As electronic components and computer technology evolved, many radars were superseded by later variants using the same basic antenna configurations. The following radars are the principle ones associated with the *Forrestal* and *Kitty Hawk* classes.

AN/SPS-8 The SPS-8 height-finding radar was carried by a number of Navy warships in the post World War II period. By the 1960s it was replaced by newer 3D radars, such as the SPS-39.

AN/SPS-10 The SPS-10 was a medium-range surface search radar with limited air search capabilities that was designed to be used on destroyers and larger ships.

AN/SPS-12 The SPS-12 air search radar entered service in the 1950s. The antenna was a truncated parabola with a massive signal horn. The range was 140 nautical miles (NM) against high altitude targets and it had some basic ECCM automatic IFF features. It was phased out in the 1980s.

AN/SPS-30 The SPS-30 was a high power air search and height finder radar used for fighter direction. It was similar in concept and function to the earlier SPS-8.

AN/SPS-37 The SPS-37 was a high power long-range 2D air search radar (up to 275 NM for large high altitude aircraft targets) that entered service in late 1960 and was installed on aircraft carriers, guided-missile cruisers, and a few guided-missile destroyers. It was succeeded by the SPS-43.

AN/SPS-39 The SPS-39 entered service in 1960 as the Navy's first three dimensional (3D) air search radar. It scanned mechanically in bearing, i.e., by rotating the antenna, and scanned vertically by emitting different frequencies at different angles from the rectangular planar antenna, which was tilted back at a fixed angle of 25 degrees. The SPS-39 had a range of 150 NM against large high altitude targets.

AN/SPS-40 The SPS-40 was a 2D air search radar that provided target detection and tracking for long-range surveillance, self defense, and fire control system designation. In the late 1980s and early 1990s, the SPS-40 was replaced by the SPS-49 radar on many ships.

AN/SPS-43 The SPS-43 was the successor to the SPS-37 and featured greatly improved ECCM performance. It entered service in 1961, but since it used an antenna nearly identical to the earlier radar, it is often difficult to distinguish them. They were usually fitted to guided-missile cruisers and aircraft carriers. It is no longer in service.

AN/SPS-48 The SPS-48 was a medium-range, 3D air search radar that could provide target position data to a ships command and control and weapons systems. It was similar in concept and appearance to the SPS-39.

AN/SPS-49 The SPS-49 is a long-range, 2D air search radar whose main function is to provide target position data to ship command and control systems. It can detect targets as high as 100,000 feet out to a distance of 300 NM.

AN/SPS-52 The SPS-52 3D air search radar can detect air targets at ranges beyond 200 NM.

AN/SPS-55 The SPS-55 was a surface search and navigation radar with a bar-shaped slotted antenna.

AN/SPS-64 The SPS-64 surface search and navigational radar has a range of about 35 NM. It has a bar-shaped slot antenna similar to many commercial marine radars.

AN/SPS-67 The SPS-67 was a replacement for the SPS-10 and originally used the same antenna. It later used a slotted antenna similar to the SPS-64.

Inboard Profile of the *Forrestal* as Built

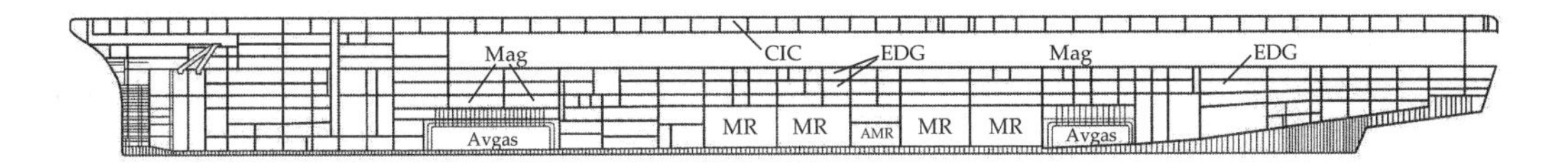

Key: MR—Machinery Room; AMR—Auxiliary Machinery Room; EDG—Engineering Diesel Generator; Avgas —Aviation Gasoline; Mag—Magazine; CIC—Combat Information Center

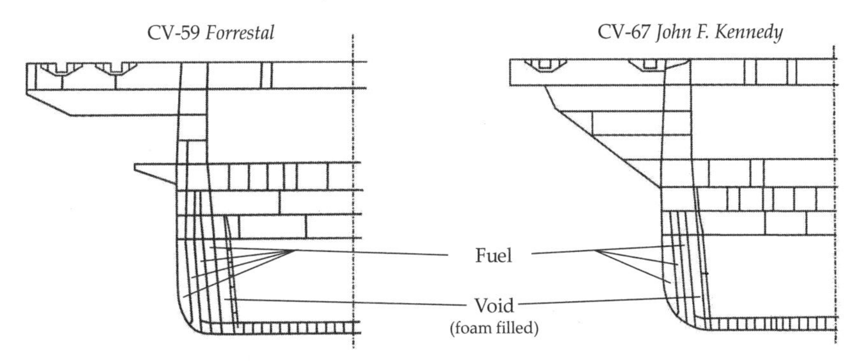

A comparison of the cross sections of the *Forrestal* and *John F. Kennedy* showing the differing internal arrangement of the side protection.

Appendix B

Individual Ships

CV-59 *Forrestal*

The motto of the *Forrestal* was "First In Defense," since she was named for James V. Forrestal, the first Secretary of Defense. *Forrestal* was assigned to the Atlantic Fleet for her entire service and made only one deployment to Vietnam. Originally home ported in Norfolk, VA, she was moved to Mayport, FL, in Sep–Oct 77. During Desert Storm, *Forrestal* was twice ordered to prepare for deployment to the war zone, but these orders were canceled both times. In 91–92 preparations began for her new role as a training carrier (AVT-59) and a move to Pensacola. She entered Philadelphia Naval Shipyard (PNSY) in Sep 92 for overhaul and conversion, but in Feb 93 the decision to decommission her was announced.

Keel Laid:	14 Jul 52
Launched:	11 Dec 54
Commissioned:	1 Oct 55
Redesignated CV-59:	30 Jun 75
Redesignated AVT-59:	4 Feb 92
Decommissioned:	11 Sep 93
Builder:	Newport News Shipbuilding Co., Newport News, VA

REMARKS: On her builders trials, the *Forrestal* experienced two main propeller thrust bearing failures and a May 56 yard period replaced the original propeller shafts, greatly improving performance. After her Jul 67 flight deck fire, emergency repairs were done at Subic Bay Jul–Aug, after which repairs were completed Sep 67–Apr 68 at Newport Naval Shipyard (NNSY). Her flight deck fire off Vietnam was not her only encounter with fire—in Jul–Aug 71 she returned to NNSY to repair a fire set by sailor on the 03 level that caused $7.5 million in damage. An Oct 89 fire caused a delay in her Med deployment and in Jun 91 she experienced a fire in an incinerator room. *Forrestal* completed SLEP between Jan 83–May 85 at PNSY.

CV-60 *Saratoga*

As the first U.S. carrier with 1,200 psi boilers, *Saratoga* suffered major boiler problems and, in general, this ship was a maintenance problem throughout her career. (She was known to some as "Sinking Sara.") The ship's emblem, the "fighting cock," was adopted to commemorate an event from the Battle of Lake Champlain during the War of 1812. At a critical point during the battle a rooster on board the corvette *Saratoga* crowed defiantly, encouraging her crew to defeat a larger British warship. Sara also at one time sported a vertical black stripe painted on her funnel as a tribute to the old *Saratoga* CV-3, which had a similar stripe to distinguish her from her sister, the *Lexington* CV-2. Assigned to the Atlantic Fleet for her entire service, her home port was Mayport. Sara made one deployment to Vietnam and participated in Desert Shield/Desert Storm.

Keel Laid:	16 Dec 52
Launched:	8 Oct 55
Commissioned:	14 Apr 56
Redesignated CV-60:	30 Jun 72
Decommissioned:	20 Aug 94
Builder:	New York Naval Shipyard, Brooklyn, NY

REMARKS: Sara completed her SLEP at PNSY Oct 80–Feb 83, after which she adopted the nickname "Super Sara."

CV-61 *Ranger*

Known as the "Top Gun of the Pacific," the *Ranger* had "Silver," a fiberglass horse, as a mascot and a crew member would dress up as the "Lone Ranger." In the all-male environment of the times, *Ranger* also had a female mannequin named "Cuddles" to serve as a morale booster on special occasions. Following her initial shakedown, *Ranger* was assigned to the Pacific Fleet and home ported in Alameda, CA. She made eight deployments to Vietnam.

Keel Laid:	2 Aug 54
Launched:	29 Sep 56
Commissioned:	10 Aug 57
Redesignated CV-61:	30 Jun 75
Decommissioned:	10 Jul 93
Builder:	Newport News Shipbuilding Co., Newport News, VA

REMARKS: *Ranger* did not receive SLEP, but instead underwent a heavy overhaul May 84–Jun 85 at Bremerton. *Ranger* was also unique in having an aluminum port aircraft elevator. She retained her forward gun sponsons throughout her career and retained some of her 5-inch guns until 1978.

CV-62 *Independence*

The *Independence* was known as "Freedom's Flagship" and was assigned to the Atlantic Fleet and home ported in Norfolk, VA. She was a participant in the naval blockade during the Cuban Missile Crisis and was the first Atlantic Fleet carrier sent to Vietnam, making one deployment May–Dec 65. Indie was to become a forward-deployed carrier home ported in Athens, Greece, in 74, but political instability in the region led to a cancellation in plans. She participated in Desert Shield but was relieved before the war broke out. In Sep 91 Indie transferred to the Pacific Fleet and replaced the *Midway* as the forward-deployed carrier in Yokosuka, Japan. In Jun 95 she was authorized to fly the "Don't Tread on Me" First Navy Jack as the oldest commissioned warship in the active fleet.

Keel Laid	1 Jul 55
Launched:	6 Jun 58
Commissioned:	10 Jan 59
Redesignated CV-62:	28 Feb 73
Decommissioned:	30 Sep 98
Builder:	New York Naval Shipyard, Brooklyn, NY

REMARKS: Indie completed her SLEP Apr 85–Jun 88 at PNSY. Her island was extensively enlarged during SLEP and other modernizations.

CV-63 *Kitty Hawk*

The *Kitty Hawk* was commonly referred to as the "Hawk" during most of her career, but adopted the nickname "Battle Cat" in 2003. Assigned to the Pacific Fleet for her entire career, she was home ported in San Diego, CA, and completed nine deployments to Vietnam. In Jul 98 she replaced the *Independence* as the forward-deployed carrier in Yokosuka, Japan. (The Hawk was authorized to fly the First Navy Jack when Indie was decommissioned.) She participated in both Enduring Freedom and Iraqi Freedom.

Keel Laid:	27 Dec 56
Christened:	21 May 60
Commissioned:	29 Apr 61
Redesignated CV-63:	29 Apr 73
Decommissioned:	12 May 09
Builder:	New York Shipbuilding Corp., Camden, NJ

REMARKS: In Aug 63 *Kitty Hawk* conducted Whale Tale carrier suitability operations of a CIA U-2; in May 64 she conducted joint operations with HMS *Victorious* (R-38) off Japan; in 1965, during first deployment to Vietnam, she experienced an engine room fire; in Dec 65 a flash fire in one of her main machinery rooms, resulting in 2 dead and 29 injured; in Jun 67 she collided with the oiler *Platte* (AO-24) during a refueling in the mid-Pacific; in Dec 67 a fire broke out on the 02 level while was moored at Subic Bay, 125 were injured with smoke inhalation; in Oct 72 she experienced racial incidents; in Dec 73 a fire in a machinery room resulted in 5 dead and 34 injured with smoke inhalation; in Jun 75 a steam line ruptured in a main machinery room, requiring repair at Subic Bay Ship Repair Facility (SRF); in Jan 83 she collided with the Canadian destroyer HCMS *Yukon*, which suffered only minor antenna damage; in Mar 84 a Soviet *Victor*-class submarine struck with the Hawk while surfacing; in Dec 91 another fire broke out in a main machinery room; in Oct–Nov 93 a damaged main engine turbine rotor had to be replaced at North Island; in Oct 00, yet another fire was caused by smoldering oil; in Sep 02 CAPT Thomas A. Hejl was relieved and in Feb 03 her new captain, CAPT Thomas A. Parker, christened the new nickname of "Battle Cat" (probably a play on her previous unofficial nickname of "Shitty Kitty," which referred to her poor materiel condition); in May–Oct 03 she underwent Drydocking Selected Restricted Availability (DSRA) at Yokosuka.

CV-64 *Constellation*

Although she was most commonly known as "Connie" the *Constellation* became "America's Flagship" following a 1981 visit from President Ronald Reagan during which he presented a presidential flag to the crew. After her initial shakedown operations in the Caribbean, she transferred to the Pacific and was home ported in San Diego. Connie was involved in the Tonkin Gulf Incident and in all made eight deployments to Vietnam. She had been scheduled to replace *Independence* in Japan in 1998 and serve through 2008, but was found to be in worse condition than *Kitty Hawk*, although she did participate in both Operations Enduring Freedom and Iraqi Freedom.

Keel Laid:	14 Sep 57
Christened:	8 Oct 60
Commissioned:	27 Oct 61
Redesignated CV-64:	30 Jun 75
Decommissioned:	7 Aug 03
Builder:	New York Naval Shipyard, Brooklyn, NY

REMARKS: Connie completed SLEP Jul 90–Mar 93 PNSY.

CV-66 *America*

Although an Atlantic Fleet carrier home ported in Norfolk, *America* made three deployments to Vietnam. She also participated in the Eldorado Canyon operations as well as Desert Shield and Desert Storm. Despite protests by former crew members and others who wanted to see "The Big A" become a memorial museum, she was scuttled 14 May 05 southeast of Cape Hatteras, after four weeks of live-fire weapons tests. *America* was the largest warship ever to be sunk.

Keel Laid:	9 Jan 61
Launched:	1 Feb 64
Commissioned:	23 Jan 65
Redesignated CV-66:	30 Jun 75
Decommissioned:	9 Aug 96
Builder:	Newport News Shipbuilding, Newport News, VA

REMARKS: *America* did not receive SLEP and was reported to be in poor condition during her last deployment. She was commissioned to reserve in Aug 96, initially laid up at Norfolk, and moved to Philadelphia in Sep 97. She was stricken for disposal in Oct 98, retroactive to Aug 96. In response to a Freedom of Information Act request from former crew members, the Navy released the exact location where *America* was sunk: 33°09′09″N 71°39′07″W. She lies upright in one piece 16,860 feet down.

CV-67 *John F. Kennedy*

John F. Kennedy is usually considered to be the only ship of her class and "Big John" was the last conventionally powered carrier built for the U.S. Navy. Originally ordered as a nuclear carrier, but converted to conventional propulsion after construction began, her island has angled funnels to direct smoke and gases away from the flight deck and she is 17 feet shorter than the *Kitty Hawk* class. The name has been adopted for the future *Gerald R. Ford*–class aircraft carrier CVN-79.

Keel Laid:	22 Oct 64
Launched:	27 May 67
Commissioned:	7 Sep 68
Redesignated CV-67:	1 Dec 74
Decommissioned:	1 Aug 07
Builder:	Newport News Shipbuilding Co., Newport News, VA

REMARKS: The *Kennedy* did not receive SLEP, but underwent Comprehensive (or Complex) Overhaul (COH), at Philadelphia Navy Yard 93–95, which involved roughly the same work. In Dec 91 a Navy Board of Inspection and Survey (INSURV) found that three of the four aircraft elevators were inoperable, two of the four catapults were in poor condition as was the flight deck firefighting equipment. The propulsion system was deemed unsafe and there was severe topside corrosion. Her captain and engineering officer were subsequently relieved.

Appendix C

Aircraft Technical Data

In 1962 Secretary of Defense Robert McNamara ordered that all services adopt a uniform system for designating military aircraft. The new system followed the previous Air Force system, as shown by the example below:

RF-4B (formerly F4H-1P under the previous Navy system)
R = modified mission (Reconnaissance)
F = basic mission (Fighter)
4 = design in mission aircraft series (4th)
B = series in design (2nd)

The table below shows old and new designations for aircraft operating from aircraft carriers.

Old	Name	New
AD	Skyraider	A-1
AJ	Savage	A-2
A3D	Skywarrior	A-3
A4D	Skyhawk	A-4
A3J	Vigilante	A-5
A2F	Intruder	A-6
FJ	Fury	F-1
F2H	Banshee	F-2
F3H	Demon	F-3
F4H	Phantom	F-4
F4D	Skyray	F-6

Old	Name	New
F8U	Crusader	F-8
F9F	Cougar	F-9
F3D	Skyknight	F-10
S2F	Tracker	S-2
WF	Tracer	E-1
W2F	Hawkeye	E-2
HUP	Retriever	H-25
HU2K	Seasprite	H-2
HSS	Sea King	H-3
HRB	Sea Knight	H-46

The following carrier aircraft are listed in approximate order of entering service:

Fixed-wing Aircraft

Grumman F9F Cougar (F9F-8)

Power Plant: One 7,250 lbs thrust Pratt & Whitney J48-P-8-A
Dimensions: Span 34 ft 6 in; length 41 ft 9 in; height 12 ft 3 in; wing area 337 sq ft
Weights: 11,866 lbs empty; 19,738 lbs gross
Performance: Max speed 647 mph at sea level; initial climb 5,750 ft/min; service ceiling 42,800 ft; range 1,200 miles
Crew: Pilot
Armament: Four 20mm cannon; four AIM-9 Sidewinder AAMs or four 500 lb bombs

Douglas AD Skyraider (AD-2)

Power Plant: One 2,700 hp Wright R-3350-26W
Dimensions: Span 50 ft 0 in; length 38 ft 2 in; height 15 ft 8 in; wing area 400 sq ft
Weights: 10,546 lbs empty; 18,263 lbs gross
Performance: Max speed 321 mph at 18,300; cruising speed 198 mph; initial climb 2,800 ft/min; service ceiling 32,700 ft; range 915 miles
Crew: Pilot
Armament: Two 20mm cannon; up to 8,000 lbs external ordnance

North American AJ Savage (AJ-2)

Power Plant: Two 2,300 hp Pratt & Whitney R-2800-44W; one 4,600 lbs thrust Allison J33-A-19
Dimensions: Span 75 ft 0 in; length 64 ft 0 in; height 21 ft 5 in; wing area 836 sq ft
Weights: 30,776 lbs empty; 48,040 lbs gross
Performance: Max speed 443 mph at 32,000 ft; cruising speed 270 mph; initial climb 930 ft/min; service ceiling 40,000 ft; range 1,714 miles.
Crew: Three
Armament: Up to 12,000 lbs internally

Douglas A3D Skywarrior (A3D-2)

Power Plant: Two 12,400 lbs thrust Pratt & Whitney J57-P-10
Dimensions: Span 72 ft 6 in; length 76 ft 4 in; height 22 ft 9.5 in; wing area 812 sq ft
Weights: 39,409 lbs empty; 82,000 lbs gross
Performance: Max speed 610 mph at 10,000; service ceiling 41,000 ft; tactical radius 1,050 miles
Crew: Three
Armament: Two 20mm cannon; up to 12,000 lbs bombs or other ordnance

Douglas A4D Skyhawk (A-4M)

Power Plant: One 11,200 lbs thrust Pratt & Whitney J52-P-408A
Dimensions: Span 27 ft 6 in; length (excluding refueling probe) 40 ft 4 in; height 15 ft 0 in; wing area 260 sq ft
Weights: 10,465 lbs empty; 24,500 lbs gross
Performance: Max speed 670 mph at sea level; initial climb 8,440 ft/min; tactical radius with 4,000 lb bomb load 340 miles
Crew: Pilot
Armament: Two 20mm cannon; up to 9,155 lbs of bombs or rockets

North American FJ Fury (FJ-4B)

Power Plant: One Wright 7,700 lbs thrust J65-W-16A
Dimensions: Span 39 ft 1 in; length 36 ft 4 in; height 13 ft 11 in; wing area 339 sq ft
Weights: 13,210 lbs empty; 20,130 lbs gross
Performance: Max speed 680 mph at 35,000 ft; rate of climb 7,660 ft/min; service ceiling 46,800 ft; range 2,020 miles (with 200 gal drop tanks and two AIM-9 Sidewinder AAMs
Crew: Pilot
Armament: Four 20mm cannon; four AIM-9 Sidewinder AAMs; up to 3,000 lbs underwing ordnance (including missiles)

Vought F7U Cutlass (F7U-3)

Power Plant: Two 4,600 lbs thrust Westinghouse J46-WE-8A
Dimensions: Span 38 ft 8 in; length 44 ft 3 in; height 14 ft 8 in; wing area 496 sq ft
Weights: 18,210 lbs empty; 31,642 lbs gross
Performance: Max speed 680 mph at 10,000 ft; initial climb 13,000 ft/min; service ceiling 40,000 ft; range 660 miles
Crew: Pilot
Armament: Four 20mm cannon; four Sparrow I AAMs

Grumman F11F Tiger (F11F-1)

Power Plant: One 7,450 lbs thrust Wright J65-W-18
Dimensions: Span 31 ft 8 in; length 46 ft 11 in; height 13 ft 3 in; wing area 250 sq ft
Weights: 13,428 lbs empty; 22,160 lbs gross
Performance: Max speed 750 mph at sea level; initial climb 5,130 ft/min; service ceiling 41,900 ft; range 1,270 miles
Crew: Pilot
Armament: Four 20mm cannon; four AIM-9 Sidewinder AAMs

McDonnell F3H Demon (F3H-2)

Power Plant: One 9,700 lbs thrust Allison J71-A-2E
Dimensions: Span 35 ft 4 in; length 58 ft 11 in; height 14 ft 7 in; wing area 519 sq ft
Weights: 22,133 lbs empty; 33,900 lbs gross
Performance: Max speed 647 mph at 30,000; initial climb 12,795 ft/min; service ceiling 42,650 ft; range 1,370 miles
Crew: Pilot
Armament: Four 20mm cannon; bombs or rockets

Douglas F4D Skyray (F4D-1)

Power Plant: One 10,200 lbs thrust (16,000 lbs with afterburner) Pratt & Whitney J57-P-8
Dimensions: Span 33 ft 6 in; length 45 ft 8 in; height 13 ft 0 in; wing area 557 sq ft
Weights: 16,024 lbs empty; 25,000 lbs gross
Performance: Max speed 695 mph at 36,000; initial climb 18,000 ft/min; service ceiling 55,000 ft; range 1,200 miles
Crew: Pilot
Armament: Four M-12 20mm cannon; four AIM-9 Sidewinders or four rocket packs

Vought F8U Crusader (F-8E)

Power Plant: One 10,700 lbs thrust (18,000 lbs with afterburner) Pratt & Whitney J57-P-20
Dimensions: Span 35 ft 8 in; length 54 ft 3 in; height 15 ft 9 in; wing area 350 sq ft
Weights: 17,836 lbs empty; 34,100 lbs gross
Performance: Max speed 1,133 mph at 35,000; initial climb 27,200 ft/min; service ceiling 52,350 ft; combat range 1,425 miles
Crew: Pilot
Armament: Four 20mm cannon; four AIM-9 Sidewinder AAMs; up to 5,000 lbs bombs or rockets

McDonnell F4H Phantom II (F-4J)

Power Plant: Two 11,870 lbs (17,900 lbs with afterburner) J79-GE-10
Dimensions: Span 38 ft 5 in; length 58 ft 3 in; height 16 ft 3 in
Weights: 30,770 lbs empty; 46,833 lbs gross
Performance: Max speed 1,584 mph at 48,000 ft; initial climb 41,250 ft/min; service ceiling 70,000 ft; range 1,956 miles (with max external fuel)
Crew: Pilot and Radar Intercept Officer
Armament: Four AIM-7 Sparrow AAMs and two to four AIM-9 Sidewinder AAMS; up to 16,000 lbs total external ordnance

North American A3J Vigilante (RA-5C)

Power Plant: Two 11,870 lbs (17,900 lbs with afterburner) J79-GE-10
Dimensions: Span 53 ft 0 in; length 75 ft 10 in; height 19 ft 5 in (15 ft 6 in with fin folded); wing area 769 sq ft
Weights: 39,600 lbs empty; 66,800 lbs normal takeoff; 80,000 lbs max
Performance: Max speed 1,385 mph; initial climb 6,600 ft/min; service ceiling 49,000 ft; combat radius 547 miles; max ferry range 2,050 miles
Crew: Pilot and Reconnaissance/Attack Navigator
Armament: Four underwing pylons (normally used for drop tanks)

Grumman A-6 Intruder (A-6E)

Power Plant: Two 9,300 lbs thrust Pratt & Whitney J52-P-8B
Dimensions: Span 53 ft 0 in; length 54 ft 7 in; height 16 ft 2 in; wing area 529 sq ft
Weights: 28,800 lbs empty; 60,626 lbs gross 64,000 lbs max
Performance: Max speed 644 mph at sea level; 474 mph cruise; service ceiling 42,400 ft; range 1,010 miles
Crew: Pilot and Bombardier/Navigator
Armament: Up to 15,000 lbs on four underwing pylons and one centerline hardpoint

LTV A-7 Corsair II (A-7E)

Power Plant: One 14,250 lbs thrust Allison TF41-A-2
Dimensions: Span 38 ft 9 in; length 46 ft 2 in; height 16 ft 1 in; wing area 375 sq ft
Weights: 19,490 lbs empty; 42,000 lbs gross
Performance: Max speed 693 mph at sea level (clean), 565 mph at 12,000 with 8,000 lbs of bombs; max climb 12,640 ft/min; max ferry range 2,300 miles
Crew: Pilot
Armament: One M61-A1 multi-barrel 20mm cannon; two AIM-9 Sidewinders; up to 10,000 lbs of bombs or rockets

Grumman WF-1 Tracer (E-1B)

Power Plant: Two 1,525 hp Wright R-1820-82A
Dimensions: Span 72 ft 4 in; length 44 ft 6 in; height 16 ft 10 in; wing area 499 sq ft
Weights: 20,638 lbs empty; 26,600 lbs gross
Performance: Max speed 227 mph at 4,000; initial climb 1,120 ft/min; service ceiling 15,800 ft; range 1,000 miles at 10,000 ft
Crew: Pilot, Copilot/Tactical Director, and two Radar Operators
Armament: None

Grumman W2F (E2) Hawkeye

Power Plant: Two 5,100 shp Allison/Rolls-Royce T56-A-427
Dimensions: Span 80 ft 7 in; length 7 ft 8.75 in; height 18 ft 3.75 in; wing area 700 sq ft
Weights: 40,200 lbs empty; 43,068 lbs loaded; 57,500 lbs max takeoff
Performance: Max speed 403 mph at sea level; cruise 295 mph; service ceiling 34,700 ft; endurance 6 hours; ferry range 1,462 miles
Crew: Pilot, Copilot, Radar Officer (RO), Combat Information Center Officer (CICO), Aircraft Control Officer (ACO)
Armament: None

Grumman F-14 Tomcat (F-14D)

Power Plant: Two 27,600 lb with afterburner General Electric F110-GE-400
Dimensions: Span 64 ft 0 in unswept (38 ft 0 in swept); length 62 ft 9 in; height 16 ft 0 in; wing area 565 sq ft
Weights: 43,735 lbs empty; 61,000 lbs loaded; 74,350 lbs max takeoff
Performance: Max speed 1,544 mph; rate of climb 4,500 ft/min; combat radius 500 miles; ferry range 1,500 miles
Crew: Pilot and Radar Intercept Officer
Armament: One M61-A1 multi-barrel 20mm cannon; six under fuselage, two under nacelle, and two wing glove hardpoints with up to 14,500 lbs ordnance (includes AIM-54 Phoenix, AIM-7 Sparrow, AIM-9 Sidewinder AAMs)

McDonnell Douglas F/A-18 Hornet (F/A18C)

Power Plant: Two 17,700 lbs thrust F404-GE-402
Dimensions: Span 40 ft 5 in; length 56 ft 0 in; height 15 ft 4 in
Weights: 23,000 lbs empty; 36,970 lbs loaded
Performance: Max speed 1,190 mph; service ceiling 50,000 ft; rate of climb 50,000 ft/min; combat radius 460 miles; range 2,070 miles
Crew: Pilot
Armament: One M61-A1 multi-barrel 20mm cannon; two wingtip missile launch rails, four under wing pylons, three under fuselage hardpoints for up to 13,700 lbs external fuel and ordnance

Boeing F/A-18E/F Super Hornet (F/A-18E)

Power Plant: Two 13,000 lbs thrust (22,000 lbs with afterburner) General Electric F414-GE-400

Dimensions: Span 44 ft 8.5 in; length 60 ft 1.25 in; height 16 ft 0 in; wing area 500 sq ft

Weights: 32,081 lbs empty; 47,000 lbs loaded; 66,000 lbs max takeoff

Performance: Max speed 1,190 mph at 40,000 ft; rate of climb 44,882 ft/min; service ceiling over 50,000 ft; combat radius 390 miles; ferry range 1,800 miles

Crew: Pilot

Armament: One M61-A1 multi-barrel 20mm cannon; two wingtip missile launch rails, six under wing pylons, three under fuselage hardpoints for up to 17,750 lbs external fuel and ordnance

Lockheed S-3 Viking (S-3A)

Power Plant: Two 9,275 lbs thrust General Electric TF34-GE-2

Dimensions: Span 68 ft 8 in; length 53 ft 4 in; height 22 ft 9 in; wing area 598 sq ft

Weights: 26,581 lbs empty; 38,192 lbs loaded; 23,831 lbs max takeoff

Performance: Max speed 493 mph at sea level; initial climb 5,120 ft/min; service ceiling 40,900 ft; range 3,182 miles; ferry range 3,875 miles

Crew: Pilot, Co-pilot, Tactical Coordinator, and Acoustic Sensor Operator

Armament: Four internal and two external hardpoints for up to 4,900 lbs ordnance

Helicopters

Piasecki HUP Retriever (HUP-2)

Power Plant: One 550 hp Continental R-975-42

Dimensions: Length 31 ft 10 in; rotor diameter 35 ft each

Weights: 5,440 lbs gross

Performance: Max speed 105 mph

Crew: Two

Kaman UH-2 Seasprite (UH-2C)

Power Plant: Two 1,250 shp General Electric GE T58-GE-8B

Dimensions: Length 52 ft 2 in; height 13 ft 6 in; rotor diameter 44 ft 0 in

Weights: 6,614 lbs empty

Performance: Max speed 137 mph

Crew: Three

Sikorsky SH-3 Sea King (SH-3H)

Power Plant: Two 1,500 shp General Electric T58-GE-10
Dimensions: Length 78 ft 8 in; rotor diameter 62 ft 0 in
Weights: 13,465 lbs empty, 18,897 lbs gross
Performance: Max speed 136 mph at sea level; cruising speed 136 mph; initial climb 2,200 ft/min; service ceiling 14,700 ft; range 625 miles
Crew: Four

Boeing Vertol H-46 Sea Knight

Power Plant: Two 1,870 shp General Electric T58-GE-16
Dimensions: Length 45 ft 8 in; height 16 ft 8.5 in; rotor diameters 51 ft 0 in
Weights: 24,300 lbs max gross takeoff
Performance: Max speed 165 mph at sea level; ceiling 10,000 ft
Crew: Four

Sikorsky H-60 (SH-60B)

Power Plant: Two 1,890 shp General Electric T700-GE-401C
Dimensions: Length 64 ft 8 in; rotor diameter 53 ft 8 in
Weights: 15,200 lbs empty, 17,758 lbs loaded for ASW
Performance: Max speed 168 mph; service ceiling 12,000 ft; range 518 miles
Crew: Three or four
Armament: Up to three Mark 46 or Mark 54 torpedoes; four AGM-114 Hellfire missiles; AGM-119 Penguin missiles; M60 or M240 machine guns; GAU-16/A or GAU-17/A Miniguns; Rapid Airborne Mine Clearance System (RAMICS) using Mark 44 Mod. 0 30mm cannon

Appendix D

Air Wings

The basic unit of naval aviation is the squadron. Different types of squadrons combine with detachments of specialty aircraft (known as "overhead") to make up carrier air wings (CVW). (When the Navy was setting up its designation system in the 1920s, the letter "V" was used to denote heavier-than-air aircraft. Thus a fighter squadron, for example, was designated "VF" followed by a number.) Initially, all the squadrons that operated from a particular ship formed her air group and the air group was named for the ship, as in "Enterprise Air Group" and all the squadrons were numbered the same as the hull number, e.g., VF-6, VB-6, etc., for the *Enterprise*, CV-6. In 1942 air groups were given numbers, the first being Carrier Air Group Nine (CAG-9), and over time air groups were assigned to different carriers as operational needs dictated. The air group designations underwent different changes in 1943 to reflect the class of aircraft carrier they were assigned to, i.e., CVBG for large carriers, CVG for carriers, CVLG for light carriers, and CVEG for the larger escort carriers, but in 1948 they all became CVGs regardless of ship class. At the time the *Forrestal* and her sisters entered service, the air groups included fighter (VF) and attack (VA) squadrons with a heavy attack squadron for nuclear strike (VAH), all-weather attack squadrons (VA[AW]) and detachments for photographic reconnaissance (VAP/VFP, and later RVAH), airborne early warning (VAW), and electronic countermeasures (VAQ), as well as helicopters (HC, HS, HU). In 1963 the air groups of attack carriers were redesignated as Carrier Air Wings (CVWs). (The aircraft assigned to Antisubmarine Warfare Carriers [CVSs] continued to be designated as Antisubmarine Air Groups [CVSGs]). The CVSG air groups were decommissioned under the "CV" concept and the VS and HS squadrons became part of the CVW air wings. Another type of air group during the 1950s was the Air Task Group (ATG), which was composed of squadrons from other air groups. Just as the air group (and later air wing) designations were no longer tied to a specific carrier, so too were the squadrons not tied to a particular air wing and changes occurred, often as new aircraft types entered service. Although an air wing might be associated with a given carrier for several years, they were on occasion reassigned, particularly when a carrier was undergoing an extensive yard period for modernization or overhaul.

The Navy, starting in World War II, had devised various symbols to identify air groups assigned to carriers. This system actually identified the specific carrier and the symbols stayed the same when a new air group was assigned. Since the symbols were hard to remember or describe over radio, the Navy adopted a letter system in July 1945. After the war the letter system began to identify the specific air group and not the carrier it was assigned to. The visual identification system had evolved to a two-letter system by late 1956. If the first letter was an "A" it indicated the air wing was assigned to the Atlantic Fleet; if the first letter was an "N" it was assigned to the Pacific Fleet. The second letter indicated the air wing. All squadrons attached to an air wing had the same two letter tail codes, but aircraft that operated with the air wing as detachments, e.g., photographic reconnaissance or helicopters, would have different tail codes. The heavy attack squadrons flying the A3D Skywarrior also had their own two letter tail codes. (Early on, squadrons assigned to the *Forrestal* had single letter tail codes.) The combination of the two letters of the tail code and the "modex" positively identify an individual aircraft within the air wing. The modex is a three-digit number used for Navy and Marine aircraft. The first number indicates the squadron and the other two a specific aircraft within a squadron. The 1xx and 2xx series are normally assigned to fighter squadrons, followed by the attack squadrons and overhead aircraft with detachment aircraft identified by one or two numbers, but there have been many combinations over the years as air wing composition has evolved. In 1955 the Navy not only adopted a new paint scheme for carrier aircraft (light gull gray overall with glossy white undersides and control surfaces) but also new squadron colors: insignia red for the first squadron, orange yellow for the second, light blue for the third, international orange for the fourth, light green for the fifth, black for the sixth, and maroon for composite squadrons. These colors often became part of the colorful markings common in the 1950s and 1960s. Each squadron would also have one aircraft nominally assigned to the commander of the air wing that would have a modex ending in double zeros and painted with all the colors of the air wing squadrons. These "CAG bird" aircraft were especially colorful and are often a subject of interest to aviation enthusiasts. (The commander of the air wing is still referred to as the CAG even though there are no longer any air groups. The CAG was usually a senior Commander who was often promoted to Captain during his tour. In 1983 Navy Secretary Lehman elevated the CAG to be co-equal with the Captain of the carrier, with both officers reporting directly to the Commander of the Carrier Battle Group. The CAG then became a "Super CAG" and a slightly junior Captain was later added as the Deputy CAG [DCAG], with the DCAG assisting the CAG until "fleeting up" to CAG. This system is still in place, but the term Super CAG soon reverted back to the traditional "CAG.")

The following is a list of attack carrier air wings and their assigned letters. There were also replacement Training Air Wings (RCVWs) and Reserve Air Wings (CVWRs). Not all of the air wings listed below were assigned to either the *Forrestal* or *Kitty Hawk*–class carriers. Also, there have been changes to the letters assigned to a particular air wing and not all air wings listed are still in commission since force levels have fluctuated over the years.

Air Wing	Tail Code
CVW-1	AB
CVW-2	NE
CVW-3	AC
RCVW-4	AD
CVW-5	NF
CVW-6	AE
CVW-7	AG
CVW-8	AJ
CVW-9	NG
CVW-10	AK

Air Wing	Tail Code
CVW-11	NH
RCVW-12	NJ
CVG-13	AE
CVW-14	NK
CVW-15	NL
CVW-17	AA
CVW-19	NM
CVWR-20	AF
CVW-21	NP
CVWR-30	ND

Deployments

The following tables show the dates of deployment, the squadrons assigned within the air group/air wing, the modex, and aircraft types. Where the tail code letters are different from the air wing, they are shown next to the modex. The abbreviations for the areas of operation are as follows:

Atlantic	Lant
Mediterranean	Med
Northern Atlantic	NorLant
Southern Atlantic	SoLant
Western Atlantic	WestLant
Caribbean	Carib
Guantánamo Bay	Gtmo

Pacific	Pac
Western Pacific	WestPac
Southern California	SoCal
Hawaiian Islands	HI
Indian Ocean	IO
Arabian/Persian Gulf	Gulf
Sea of Japan	SOJ

CV-59 *Forrestal*

Deployment Dates	Squadrons	Modex	Aircraft Type	Air Wing/ Tail Code	Area of Operations	Operations/ Exercises
Jan–Mar 56	VF-41	100	F2H-3	ATG-181 I	Gtmo	Shakedown
	VF-21	200	FJ-3			
	VA-86	300	F7U-3M			
	VA-42	400	AD-6			
	VAH-7 Det.42	NHx	AJ-2			
	VC-12 Det.42	NE7xx	AD-5W			
	VC-33 Det.42	SS8xx	AD-5N			
	VC-62 Det.42	PL9xx	F2H-2P			
	HU-2 Det.42	URxx	HUP-2			
Nov–Dec 56	VF-171	R100	F2H-3/4	CVG-1 T	Azores	Suez Crisis
	VA-76	O300	F9F-8/8B			
	VA-15	400	AD-6			
	VAW-12 Det.42	NE7xx	AD-5W			
	VA(AW)-33 Det.42	SS8xx	AD-5N			
	VFP-62 Det.42	PL9xx	F2H-2P			
	HU-2 Det.42	URxx	HUP-22			
Jan–Jul 57	VAH-1	TBx	A3D-1	CVG-1 T	Med	
	VF-14	100	F3H-2N			
	VF-84	O200	FJ-3M			
	VA-76	O300	F9F-8B			
	VA-15	400	AD-6			
	VAW-12 Det.42	NE7xx	AD-5W			
	VA(AW)-33 Det.42	SS8xx	AD-5N			
	VFP-62 Det.42	9xx	F2H-2P			
	HU-2 Det.42	HUxx	HUP-2			
Aug–Oct 57	VAH-1	12-Jan	A3D-1	CVG-1 AB	NorLant	Strikeback
	VF-14	100	F3H-2N			
	VF-84	200	FJ-3M			
	VA-15	400	AD-6			
	VAW-12 Det.42	7xx	AD-5W			
	VA(AW)-33 Det.42	GD80x	AD-5N			
	HU-2 Det.42.	HUxx	HUP-2			
Mar–May 58	VF-82	100	F3H-2/2N	CVG-8 AJ	NC, GA, FL coasts	
	VF-21	AF200	F11F-1			
	VA-72	AG300	A4D-2			
	VA-85	500	AD-6			
	VAW-12 Det.42	GE7xx	AD-5W			
	VFP-62 Det.42	GA9xx	F9F-8P			
	HU-2 Det.42	HUxx	HUP-2			
Jul–Aug 58	VF-102	100	F4D-1	CVG-10 AK	Med	Lebanon Crisis
	VF-103	200	F8U-1			
	VA-12	300	A4D-2			
	VA-104	400	AD-6			
	VAH-5	500	A3D-2			
	VAW-12 Det.42	7xx	AD-5W			
	VA(AW)-33 Det.42	8xx	AD-5N			
	VFP-62 Det.42	9xx	F8U-1P			
	HU-2 Det.42	HUxx	HUP-2			

Deployment Dates	Squadrons	Modex	Aircraft Type	Air Wing/ Tail Code	Area of Operations	Operations/ Exercises
Sep 58–Mar 59	VF-102	100	F4D-1	CVG-10 AK	Med	
	VF-103	200	F8U-1			
	VA-12	300	A4D-2			
	VA-104	400	AD-6			
	VAH-5	500	A3D-2			
	VAW-12 Det.42	72x	AD-5W			
	VA(AW)-33 Det.42	8xx	AD-5N			
	VFP-62 Det.42	9xx	F8U-1P			
	HU-2 Det.42	HUxx	HUP-2			
Jan–Aug 60	VF-102	100	F4D-1	CVG-8 AJ	Med	
	VF-103	200	F8U-2			
	VA-83	300	A4D-2			
	VA-81	400	A4D-2			
	VA-85	500	AD-6			
	VAH-5	600	A3D-2			
	VAW-12 Det.42	7xx	AD-5W			
	VAW-33 Det.42	GD8xx	AD-5Q			
	VFP-62 Det.42	GA9xx	F8U-1P			
	HU-2 Det.42	HUxx	HUP-2			
Feb–Aug 61	VF-102	100	F4D-1	CVG-8 AJ	Med	
	VF-103	200	F8U-2			
	VA-83	300	A4D-2N			
	VA-81	400	A4D-2			
	VA-85	500	AD-6			
	VAH-5	600	A3D-2			
	VAW-12 Det.42	7xx	WF-2			
	VAW-33 Det.42	8xx	AD-5Q			
	VFP-62 Det.42	GA9xx	F8U-1P			
	HU-2 Det.42	HUxx	HUP-3			
Jan–Feb 62	VF-74	100	F4H-1	CVG-8 AJ	Carib	
	VF-103	200	F8U-2			
	VA-83	300	A4D-2N			
	VA-81	400	A4D-2			
	VA-85	500	AD-6			
	VAH-5	600	A3D-2			
	VAW-12 Det.42	GE7xx	WF-2			
	VAW-33 Det.42	GD8xx	AD-5Q			
	VFP-62 Det.42	GA9xx	F8U-1P			
	HU-2 Det.42	HUxx	HUP-3			
Apr 62	VF-74	100	F4H-1	CVG-8 AJ	Carib	
	VF-103	200	F8U-2			
	VA-83	300	A4D-2N			
	VA-81	400	A4D-2			
	VA-85	500	AD-6			
	VAH-5	600	A3D-2			
	VAW-12 Det.42	GE7xx	WF-2			
	VAW-33 Det.42	GD8xx	AD-5Q			
	VFP-62 Det.42	GA9xx	F8U-1P			
	HU-2 Det.42	HUxx	HUP-3			

CV-59 *Forrestal* (continued)

Deployment Dates	Squadrons	Modex	Aircraft Type	Air Wing/ Tail Code	Area of Operations	Operations/ Exercises
Jun–Jul 62	VF-74	100	F4H-1	CVG-8 AJ	Carib	
	VF-103	200	F8U-2			
	VA-83	300	A4D-2N			
	VA-81	400	A4D-2			
	VA-85	500	AD-6			
	VAH-5	600	A3D-2			
	VAW-12 Det.42	GE7xx	WF-2			
	VAW-33 Det.42	GD8xx	AD-5Q			
	VFP-62 Det.42	GA9xx	F8U-1P			
	HU-2 Det.42	HUxx	HUP-3			
Aug 62–Mar 63	VF-74	100	F4H-1	CVG-8 AJ	Med	
	VF-103	200	F8U-2			
	VA-83	300	A4D-2N			
	VA-81	400	A4D-2			
	VA-85	500	AD-6			
	VAH-5	600	A3D-2			
	VAH-11 Det.59	AB610	A3D-2			
	VAW-12 Det.59	7xx	WF-2			
	VAW-33 Det.59	8xx	AD-5Q			
	VFP-62 Det.59	9xx	F8U-1P			
	HU-2 Det.59	HUxx	HUP-3			
Apr–May 64	VF-74	100	F-4B	CVW-8 AJ	WestLant	
	VF-103	200	F-8C			
	VA-83	300	A-4E			
	VA-81	400	A-4E			
	VA-85	500	A-6A			
	VAW-12 Det.59	7xx	E-1B			
	VAW-33 Det.59	8xx	EA-1F			
	VFP-62 Det.59	9xx	RF-8A			
	HS-5 Det.59	AT5x-6x	SH-3A			
Jul 64–Mar 65	VF-74	100	F-4B	CVW-8 AJ	Med	
	VF-103	200	F-8E			
	VA-83	300	A-4E			
	VA-81	400	A-4E			
	VMA-331	500	A-4E			
	VAH-6	600	A-3B			
	VAW-12 Det.59	700	E-1B			
	VAW-33 Det.59	8xx	EA-1F			
	VFP-62 Det.59	9xx	RF-8A			
	HU-2 Det.59	HUxx	UH-2A/B			
Aug 65–Apr 66	VF-74	100	F-4B	CVW-8 AJ	Med	
	VMF(AW)-451	200	F-8D			
	VA-83	300	A-4E			
	VA-81	400	A-4E			
	VA-112	500	A-4C			
	VAH-11	600	A-3B			
	VAW-12 Det.59	7xx	E-1B			
	VFP-62 Det.59	9xx	RF-8A			
	HC-2 Det.59	HU xx	UH-2A/B			

Deployment Dates	Squadrons	Modex	Aircraft Type	Air Wing/ Tail Code	Area of Operations	Operations/ Exercises
Jun–Sep 67	VF-11	100	F-4B	CVW-17 AA	WestPac Vietnam	
	VF-74	200	F-4B			
	VA-106	300	A-4E			
	VA-46	400	A-4E			
	VA-65	500	A-6A			
	RVAH-11	600	RA-5C			
	VAW-123	730	E-2A			
	VAH-10 Det.59	810	KA-3B			
	VAW-33 Det.59	90x	EA-1F			
	HC-2 Det.59	HUxx	UH-2A			
Jul 68–Apr 69	VF-11	100	F-4B	CVW-17 AA	Med	
	VF-74	200	F-4B			
	VA-34 (*1)	300	A-4C			
	VA-15 (*1)	400	A-4E			
	VA-152	500	A-4C			
	RVAH-12	600	RA-5C			
	VAH-10 Det.59	610	KA-3B/EKA-3B			
	VAW-123	730	E-2A			
	HC-2 Det.59	HUxx	SH-3D			
Dec 69–Jul 70	VF-11	100	F-4B	CVW-17 AA	Med	
	VF-74	200	F-4B			
	VA-66	300	A-4C			
	VA-216	400	A-4E			
	VA-36	500	A-4C			
	RVAH-13	600	RA-5C			
	VAH-10 Det.59	610	KA-3B			
	VAW-126	760	E-2A			
	HS-11	xxx	SH-3D			
Jan–Jul 71	VF-11	100	F-4B	CVW-17 AA	Med	
	VF-74	200	F-4B			
	VA-83	300	A-7E			
	VA-81	400	A-7E			
	VA-85	500	A-6A/KA-6D			
	RVAH-7	600	RA-5C			
	VMCJ-2 Det.A59	610	EA-6A			
	VAW-126	760	E-2B			
	HS-3	00x	SH-3D			
Sep 72–Jul 73	VF-11	100	F-4B	CVW-17 AA	Med	
	VMFA-531	200	F-4B			
	VA-83	300	A-7E			
	VA-81	400	A-7E			
	VA-85	500	A-6E/KA-6D			
	RVAH-9	600	RA-5C			
	VAQ-135 Det.2	610	EKA-3B			
	VAW-126	760	E-2B			
	HS-3	0	SH-3D			

CV-59 *Forrestal* (continued)

Deployment Dates	Squadrons	Modex	Aircraft Type	Air Wing/ Tail Code	Area of Operations	Operations/ Exercises
Mar–Sep 74	VF-11	100	F-4J	CVW-17 AA	Med	
	VF-74	200	F-4J			
	VA-83	300	A-7E			
	VA-81	400	A-7E			
	VA-85	500	A-6E/KA-6D			
	RVAH-6	600	RA-5C			
	VAW-126	760	E-2B			
	HS-3	00x	SH-3D			
Mar–Sep 75	VF-11	100	F-4J	CVW-17 AA	Med	
	VF-74	200	F-4J			
	VA-83	300	A-7E			
	VA-81	400	A-7E			
	VA-85	500	A-6E/KA-6D			
	RVAH-7	600	RA-5C			
	VAQ-134	610	EA-6B			
	VAW-111	700	E-2B			
	HS-3	00x	SH-3D			
1976–77	VF-11	100	F-4J	CVW-17	(Not deployed)	
	VMFA-451	200	F-4J			
	VA-83	300	A-7E			
	VA-81	400	A-7E			
	VA-85	500	A-6E/KA-6D			
	RVAH-12	600	RA-5C			
	VAQ-	xxx	EA-6B			
	VAW-	xxx	E-2			
	HS-3	xxx	SH-3D			
Apr–Oct 78	VF-11	100	F-4J	CVW-17 AA	Med NorLant	Dawn Patrol Northern Wedding Display Determination Windbreak
	VF-74	200	F-4J			
	VA-83	300	A-7E			
	VA-81	400	A-7E			
	VA-85	500	A-6E/KA-6D			
	VFP-63 Det.	600	RF-8G			
	VAQ-130	610	EA-6B			
	VS-30	700	S-3A			
	HS-3	730	SH-3D			
	VAW-116	10	E-2B			
Nov 79–May 80	VF-11	100	F-4J	CVW-17 AA	Med	
	VF-74	200	F-4J			
	VA-83	300	A-7E			
	VA-81	400	A-7E			
	VA-85	500	A-6E/KA-6D			
	VFP-63 Det.	600	RF-8G			
	VAQ-133	610	EA-6B			
	VS-30	700	S-3A			
	HS-3	730	SH-3D/H			
	VAW-125	10	E-2C			

Deployment Dates	Squadrons	Modex	Aircraft Type	Air Wing/ Tail Code	Area of Operations	Operations/ Exercises
Mar–Sep 81	VMFA-115	100	F-4J	CVW-17 AA	NorLant Med	Ocean Venture '81
	VF-74	200	F-4J			
	VA-83	300	A-7E			
	VA-81	400	A-7E			
	VA-85	500	A-6E/KA-6D			
	VAW-125	600	E-2C			
	VAQ-130	610	EA-6B			
	VS-30	700	S-3A			
	HS-3	730	SH-3H			
Jun–Nov 82	VF-74	100	F-4S	CVW-17 AA	Med IO	
	VF-103	200	F-4S			
	VA-83	300	A-7E			
	VA-81	400	A-7E			
	VA-85	500	A-6E/KA-6D			
	VAW-125	600-603	E-2C			
	VAQ-130	604-607	EA-6B			
	HS-3	610	SH-3H			
	VS-30	700	S-3A			
Jul–Aug 85	VF-11	100	F-14A	CVW-6 AE	Carib	
	VF-31	200	F-14A			
	VA-15	300	A-7E			
	VA-87	400	A-7E			
	VA-176	500	A-6E/KA-6D			
	VAW-122	600-603	E-2C			
	VAQ-131	604-607	EA-6B			
	HS-15	610	SH-3H			
	VS-28	700	S-3A			
Mar–Apr 86	VF-11	100	F-14A	CVW-6 AE	WestLant	
	VF-31	200	F-14A			
	VA-37	300	A-7E			
	VA-105	400	A-7E			
	VA-176	500	A-6E/KA-6D			
	VAW-122	600-603	E-2C			
	VAQ-132	604-607	EA-6B			
	HS-15	610	SH-3H			
	VS-28	700	S-3A			
Jun–Nov 86	VF-11	100	F-14A	CVW-6 AE	Med	Display Determination
	VF-31	200	F-14A			
	VA-37	300	A-7E			
	VA-105	400	A-7E			
	VA-176	500	A-6E/KA-6D			
	VAW-122	600-603	E-2C			
	VAQ-132	604-607	EA-6B			
	HS-15	610	SH-3H			
	VS-28	700	S-3A			

CV-59 *Forrestal* (continued)

Deployment Dates	Squadrons	Modex	Aircraft Type	Air Wing/ Tail Code	Area of Operations	Operations/ Exercises
Jun 87	VF-201	100	F-14A	CVWR-20 AF	WestLant	
	VF-202	200	F-14A			
	VA-203	300	A-7E			
	VA-204	400	A-7E			
	VA-205	500	A-7E			
	VFP-206	600	RF-8G			
	VAK-208	610	KA-3B			
	HS-75	xxx	SH-3D			
	VAQ-209	xxx	EA-6A			
	VAW-78	xxx	E-2C			
Aug–Oct 87	VF-11	100	F-14A	CVW-6 AE	NorLant	Ocean Safari 87
	VF-31	200	F-14A			
	VA-37	300	A-7E			
	VA-105	400	A-7E			
	VA-176	500	A-6E/KA-6D			
	VAW-122	600-603	E-2C			
	VAQ-132	604-607	EA-6B			
	HS-15	610	SH-3H			
	VS-28	700	S-3A			
Apr–Oct 88	VF-11	100	F-14A	CVW-6 AE	Med IO NorLant	Earnest Will
	VF-31	200	F-14A			
	VA-37	300	A-7E			
	VA-105	400	A-7E			
	VA-176	500	A-6E/KA-6D			
	VAW-122	600-603	E-2C			
	VAQ-132	604-607	EA-6B			
	HS-15	610	SH-3H			
	VS-28	700	S-3A			
Nov 89–Apr 90	VF-11	100	F-14A	CVW-6 AE	Med	Harmonie Sud Tunisian Amphibious National Week
	VF-31	200	F-14A			
	VA-37	300	A-7E			
	VA-105	400	A-7E			
	VA-176	500	A-6E/KA-6D			
	VAW-122	600-603	E-2C			
	VAQ-142	604-607	EA-6B			
	HS-15	610	SH-3H			
	VS-28	700	S-3A			
May–Dec 91	VF-11	100	F-14A	CVW-6 AE	Med	Provide Comfort
	VF-31	200	F-14A			
	VFA-132	300	FA-18C			
	VFA-137	400	FA-18C			
	VA-176	500	A-6E/KA-6D			
	VAW-122	600	E-2C			
	HS-15	610	SH-3H			
	VAQ-133	620	EA-6B			
	VS-28	700	S-3B			

CV-60 *Saratoga*

Deployment Dates	Squadrons	Modex	Aircraft Type	Air Wing/ Tail Code	Area of Operations	Operations/ Exercises
Aug–Oct 56	VF-22	100	F2H-4	CVG-4 F		Shakedown
	VF-43	200	F9F-8/8B			
	VA-44	400	F9F-8			
	VA-45	500	AD-6			
	VMF(N)-533	Alxx	F2H-4			
	HU-2 Det.43	URxx	HUP-2			
Sep–Oct 57	VF-101	AP100	F4D-1	CVG-7 AG	NorLant	Strikeback
	VF-61	200	F3H-2M			
	VA-72	300	A4D-1			
	VA-75	500	AD-6			
	VAH-9 Det.43	GMx	A3D-2			
	VA(AW)-33 Det.43	GD8xx	AD-5N			
	VFP-62 Det.43	GA9xx	F2H-2P			
	HU-2 Det.43	HUxx	HUP-2			
Feb–Oct 58	VF-31	100	F3H-2N	CVG-3 AC	Med	
	VF-32	200	F8U-1			
	VA-34	300	A4D-1			
	VA-35	400	AD-6			
	VAH-9	xx	A3D-2			
	VAW-12 Det.43	7xx	AD-5W			
	VA(AW)-33 Det.43	85x	AD-5N			
	VFP-62 Det.43	96x	F9F-8P			
	HU-2 Det.43	HUxx	HUP-2			
May–Jun 59	VF-31	100	F3H-2	CVG-3 AC	WestLant	
	VF-32	200	F8U-1			
	VA-34	300	A4D-2			
	VA-35	400	AD-6			
	VAH-9	500	A3D-2			
	VA-36	600	A4D-2			
	VA(AW)-33 Det.43	GD8xx	AD-5N/Q			
	VFP-62 Det.43	GA9xx	F8U-1P			
	HU-2 Det.43	HUxx	HUP-2			
Aug 59–Feb 60	VF-31	100	F3H-2	CVG-3 AC	Med	
	VF-32	200	F8U-1E			
	VA-34	300	A4D-2			
	VA-35	400	AD-6			
	VAH-9	500	A3D-2			
	VA-36	600	A4D-2			
	VAW-12 Det.43	GE7xx	AD-5W			
	VAW-33 Det.43	GD8xx	AD-5Q			
	VFP-62 Det.43	9xx	F8U-1P			
	HU-2 Det.43	HUxx	HUP-2			

CV-60 *Saratoga* (continued)

Deployment Dates	Squadrons	Modex	Aircraft Type	Air Wing/ Tail Code	Area of Operations	Operations/ Exercises
Aug 60–Feb 61	VF-31	100	F3H-2	CVG-3 AC	Med NorLant	(VMA-224 Aug-Sep 60)
	VF-32	200	F8U-1/1E/2			
	VA-34	300	A4D-2			
	VA-35	400	AD-6			
	VAH-9	500	A3D-2			
	VA-36	600	A4D-2			
	VAW-12 Det.43	700	WF-2			
	VAW-33 Det.43	800	AD-5Q			
	VFP-62 Det.43	9xx	F8U-1P			
	HU-2 Det.43	HUxx	HUP-2/3			
	VMA-224	WKxx	A4D-2			
Jul–Aug 61	VF-31	100	F3H-2	CVG-3 AC	Carib	
	VF-32	200	F8U-2N			
	VA-34	300	A4D-2			
	VA-35	400	AD-6			
	VAH-9	500	A3D-2			
	VA-36	600	A4D-2N			
	VAW-12 Det.43	GE7xx	WF-2			
	VAW-33 Det.43	GD8xx	AD-5Q			
	VFP-62 Det.43	GA9xx	F8U-1P			
	HU-2 Det.43	HUxx	HUP-3			
Nov 61–May 62	VF-31	100	F3H-2	CVG-3 AC	Med	
	VF-32	200	F8U-2N			
	VA-34	300	A4D-2			
	VA-35	400	AD-6			
	VAH-9	500	A3D-2			
	VA-36	600	A4D-2N			
	VAW-12 Det.43	700	WF-2			
	VAW-33 Det.43	800	AD-5Q			
	VFP-62 Det.43	9xx	F8U-1P			
	HU-2 Det.43	HUxx	HUP-3			
Dec 62	VF-32	200	F-8D	CVG-3 AC	Carib	Cuban Missile Crisis
	VA-35	400	A-1H/A-1E			
	VA-36	600	A-4C			
	VAW-12 Det.60	GE7xx	E-1B			
	VAW-33 Det.60	GD8xx	EA-1F			
	VFP-62 Det.60	GA9xx	RF-8A			
	HU-2 Det.60	HUxx	UH-25B			
Mar–Oct 63	VF-31	100	F-3B	CVG-3 AC	Med	
	VF-32	200	F-8D			
	VA-34	300	A-4C			
	VA-35	400	A-1H/A-1E			
	VAH-9	500	A-3B			
	VA-36	600	A-4C			
	VAW-12 Det.60	700	E-1B			
	VAW-33 Det.43	800	AD-5Q			
	VFP-62 Det.60	920	RF-8A			
	HU-2 Det.60	HUxx	UH-25B			

Deployment Dates	Squadrons	Modex	Aircraft Type	Air Wing/ Tail Code	Area of Operations	Operations/ Exercises
Nov 64–Jul 65	VF-31	100	F-4B	CVW-3 AC	Med	
	VF-32	200	F-8D			
	VA-34	300	A-4C			
	VA-35	400	A-1H			
	RVAH-9	500	RA-5C			
	VA-36	600	A-4C			
	VAW-12 Det.60	7xx	E-1B			
	VAH-10 Det.60	800	A-3B			
	HU-2 Det.60	HUxx	UH-2A			
Jan–Feb 66	VF-31	100	F-4B	CVW-3 AC	WestLant	
	VF-103	200	F-4B			
	VA-34	300	A-4C			
	VA-106	400	A-4C			
	VA-46	500	A-4C			
	RVAH-12	600	RA-5C			
	VAW-12 Det.60	7xx	E-1B			
	VAH-10 Det.60	800	A-3B			
	HC-2 Det.60	HUxx	UH-2A/B			
Mar–Oct 66	VF-31	100	F-4B	CVW-3 AC	Med	
	VF-103	200	F-4B			
	VA-34	300	A-4C			
	VA-106	400	A-4C			
	VA-46	500	A-4C			
	RVAH-12	600	RA-5C			
	VAW-12 Det.60	7xx	E-1B			
	VAH-10 Det.60	800	A-3B			
	HC-2 Det.60	HUxx	UH-2B			
May–Dec 67	VF-31	100	F-4B	CVW-3 AC	Med	
	VF-103	200	F-4B			
	VA-176	400	A-1H			
	VA-216	600	A-4B			
	VAW-121 Det.60	7xx	E-1B			
	VAH-10 Det.60	800	KA-3B			
	RVAH-9	900	RA-5C			
	HC-2 Det.60	HUxx	UH-2A			
Jul 68–Jan 69	VF-31	100	F-4J	CVW-3 AC	Med	
	VF-103	200	F-4J			
	VA-46	300	A-7B			
	VA-75	500	A-6A			
	RVAH-1	600	RA-5C			
	VAH-10 Det.60	610	EKA-3B			
	VAW-125	750	E-2A			
	HC-2 Det.60	HUxx	UH-2C			

CV-60 *Saratoga* (continued)

Deployment Dates	Squadrons	Modex	Aircraft Type	Air Wing/ Tail Code	Area of Operations	Operations/ Exercises
May–Jun 69	VF-31	100	F-4J	CVW-3 AC	WestLant	
	VF-103	200	F-4J			
	VA-46	300	A-7B			
	VA-113	400	A-7B			
	VA-75	500	A-6A			
	RVAH-1	600	RA-5C			
	VAH-10 Det.60	610	EKA-3B			
	VAW-125	750	E-2A			
	HC-2 Det.60	HUxx	UH-2C			
Jul 69–Jan 70	VF-31	100	F-4J	CVW-3 AC	Med	
	VF-103	200	F-4J			
	VA-46	300	A-7B			
	VA-113	400	A-7B			
	VA-75	500	A-6A			
	RVAH-1	600	RA-5C			
	VAH-10 Det.60	610	EKA-3B			
	VAW-125	750	E-2A			
	HC-2 Det.60	HUxx	HH-2D			
Jun–Nov 70	VF-31	100	F-4J	CVW-3 AC	Med	
	VF-103	200	F-4J			
	VA-37	300	A-7A			
	VA-105	400	A-7A			
	VA-75	500	A-6A			
	RVAH-9	600	RA-5C			
	VAQ-130 Det.60/4	610	EKA-3B			
	VAW-123	730	E-2B			
	HC-2 Det.60	HUxx	HH-2D			
Jun–Oct 71	VF-31	100	F-4J	CVW-3 AC	NorLant Med	Magic Sword II
	VF-103	200	F-4J			
	VA-37	300	A-7A			
	VA-105	400	A-7A			
	VA-75	500	A-6A/B/KA-6D			
	RVAH-9	600	RA-5C			
	VMCJ-2 Det.A60	610	EA-6A			
	VAW-123	730	E-2B			
	VS-28	220	S-2E			
	HS-7	550	SH-3D			
Apr 72–Feb 73	VF-31	100	F-4J	CVW-3 AC	SoLant IO WestPac Vietnam	
	VF-103	200	F-4J			
	VA-37	300	A-7A			
	VA-105	400	A-7A			
	VA-75	500	A-6A/KA-6D			
	RVAH-1	600	RA-5C			
	VMCJ-2 Det.1	CY610	EA-6A			
	VAW-123	10	E-2B			
	HS-7	550	SH-3D			

Deployment Dates	Squadrons	Modex	Aircraft Type	Air Wing/ Tail Code	Area of Operations	Operations/ Exercises
Sep 74–Mar 75	VF-31	100	F-4J	CVW-3 AC	Med	
	VF-103	200	F-4J			
	VA-37	300	A-7E			
	VA-105	400	A-7E			
	VA-75	500	A-6E/KA-6D			
	RVAH-11	600	RA-5C			
	VS-24	610	S-2G			
	HS-7	700	SH-3H			
	VAW-123	720	E-2C			
Jan–Jul 76	VF-31	100	F-4J	CVW-3 AC	Med	
	VF-103	200	F-4J			
	VA-37	300	A-7E			
	VA-105	400	A-7E			
	VA-75	500	A-6E/KA-6D			
	VFP-63 Det.3	600-603	RF-8G			
	VAQ-131	604-607	EA-6B			
	VS-22	610-621	S-3A			
	VAW-123	10	E-2C			
	HS-7	1	SH-3H			
Jul–Dec 77	VF-31	100	F-4J	CVW-3 AC	Med	
	VF-103	200	F-4J			
	VA-37	300	A-7E			
	VA-105	400	A-7E			
	VA-75	500	A-6E/KA-6D			
	VAQ-138	600	EA-6B			
	VS-22	610-621	S-3A			
	VAW-123	10	E-2C			
	HS-7	1	SH-3H			
Oct 78–Apr 79	VF-31	100	F-4J	CVW-3 AC	Med	
	VF-103	200	F-4J			
	VA-37	300	A-7E			
	VA-105	400	A-7E			
	VA-75	500	A-6E/KA-6D			
	RVAH-12	601-604	RA-5C			
	VAQ-136	614-617	EA-6B			
	VS-22	700	S-3A			
	HS-7	730	SH-3H			
	VAW-123	10	E-2C			
Mar–Aug 80	VF-31	100	F-4J	CVW-3 AC	Med	
	VF-103	200	F-4J			
	VA-37	300	A-7E			
	VA-105	400	A-7E			
	VA-75	500	A-6E/KA-6D			
	VAW-123	600	E-2C			
	VS-22	700	S-3A			
	HS-7	1	SH-3H			

CV-60 *Saratoga* (continued)

Deployment Dates	Squadrons	Modex	Aircraft Type	Air Wing/ Tail Code	Area of Operations	Operations/ Exercises
Apr–Oct 84	VF-74	100	F-14A	CVW-17 AA	Med	
	VF-103	200	F-14A			
	VA-83	300	A-7E			
	VA-81	400	A-7E			
	VMA(AW)-533	500	A-6E/KA-6D			
	VAW-125	600-603	E-2C			
	VMAQ-2 Det.X	604-607	EA-6B			
	HS-3	610	SH-3H			
	VS-30	700	S-3A			
	VQ-2 Det.B	JQxx	EA-3B			
Aug 85–Apr 86	VF-74	100	F-14A	CVW-17 AA	Med IO	
	VF-103	200	F-14A			
	VA-83	300	A-7E			
	VA-81	400	A-7E			
	VA-85	500	A-6E/KA-6D			
	VAW-125	600-603	E-2C			
	VAQ-137	604-607	EA-6B			
	HS-3	610	SH-3H			
	VS-30	700	S-3A			
	VQ-2 Det.	JQxx	EA-3B			
Jun–Nov 87	VF-74	100	F-14A	CVW-17 AA	Med	
	VF-103	200	F-14A			
	VA-83	300	A-7E			
	VA-81	400	A-7E			
	VA-85	500	A-6E/KA-6D			
	VAW-125	600-603	E-2C			
	VAQ-137	604-607	EA-6B			
	HS-3	610	SH-3H			
	VS-30	700	S-3A			
	VQ-2 Det.	JQ0xx	EA-3B			
Aug 90–Mar 91	VF-74	100	F-14B	CVW-17 AA	Med Red Sea	Desert Shield Desert Storm
	VF-103	200	F-14B			
	VA-81	300	FA-18C			
	VA-83	400	FA-18C			
	VA-35	500	A-6E/KA-6D			
	VAW-125	600-603	E-2C			
	VAQ-132	604-607	EA-6B			
	VS-30	610	SH-3H			
	HS-3	700	S-3B			

Deployment Dates	Squadrons	Modex	Aircraft Type	Air Wing/ Tail Code	Area of Operations	Operations/ Exercises
May–Nov 92	VF-74	100	F-14B	CVW-17 AA	Med IO	Southern Watch Display Determination '92
	VF-103	200	F-14B			
	VFA-83	300	FA-18C			
	VFA-81	400	FA-18C			
	VA-35	500	A-6E/KA-6D			
	VAW-125	600	E-2C			
	HS-9 (*2)	610	SH-3H			
	VAQ-132	620	EA-6B			
	VS-30	700	S-3B			
	VX-1 Det.	JAxx	ES-3A			
	VRC-40 Det.	24, xx	C-2A			
Jan–Jun 94	VF-103	200	F-14B	CVW-17 AA	Med IO	Deny Flight Provide Promise Dynamic Impact '94 Iles D'Or
	VFA-83	300	FA-18C			
	VFA-81	400	FA-18C			
	VA-35	500	A-6E/KA-6D			
	VAW-125	600	E-2C			
	HS-15	610	SH-60F/HH-60H			
	VAQ-132	620	EA-6B			
	VS-30	700	S-3B			
	VQ-6 Det.A	760	ES-3A			

CV-61 *Ranger*

Deployment Dates	Squadrons	Modex	Aircraft Type	Air Wing/ Tail Code	Area of Operations	Operations/ Exercises
Oct–Dec 57	VF-82	100	F3H-2/2N	CVG-8 AJ	SoLant	Shakedown
	VF-21	AM200	F11F-1			
	VA-12	AB300	A4D-1			
	VA-85	500	AD-6/5			
May 58	VF-144	300	F9F-8	CVG-14 NK	WestLant	
	VA-116	400	FJ-4B			
	VA-145	500	AD-6/5			
	VA-146	600	FJ-4B			
	VAH-6	ZCxx	A3D-2			
	VAW-11 Det.F	RR7xx	AD-5W			
	VA(AW)-35 Det.F	VV8xx	AD-5N			
	VFP-61 Det.F	PP98x	F9F-8P			
	HU-1 Det.F	UPxx	HUP-2			
Jun–Aug 58	VF-144	300	F9F-8	CVG-14 NK	SoLant SoPac	
	VA-116	400	FJ-4B			
	VA-145	500	AD-6			
	VA-146	600	FJ-4B			
	VAW-11 Det.F	RR7xx	AD-5W			
	VA(AW)-35 Det.F	VV8xx	AD-5N			
	VFP-61 Det.F	PP98x	F9F-8P			
	HU-1 Det.F	UPxx	HUP-2			
Nov 58	VF-142	100	F8U-1	CVG-14 NK	EastPac	
	VF-141	200	F4D-1			
	VA-146	300	FJ-4B			
	VA-116	400	FJ-4B			
	VA-145	500	AD-6/5			
	VAH-6	ZCxx	A3D-2			
	VAW-11 Det.F	RR7xx	AD-5W			
	VA(AW)-35 Det.F	VV8xx	AD-5N			
	VFP-61 Det.F	PP9xx	F8U-1P			
	HU-1 Det.F	UPxx	HUP-2			
Jan–Jul 59	VF-142	100	F8U-1	CVG-14 NK	WestPac	
	VF-141	200	F4D-1			
	VA-146	300	FJ-4B			
	VA-116	400	FJ-4B			
	VA-145	500	AD-6/5			
	VAH-6	ZCxx	A3D-2			
	VAW-11 Det.FN	RR7xx	AD-5N			
	VAW-11 Det.FW	RR7xx	AD-5W/5			
	VA(AW)-35 Det.F	VV8xx	AD-5N			
	VFP-61 Det.F	PP9xx	F8U-1P			
	HU-1 Det.F	UPxx	HUP-2			

Deployment Dates	Squadrons	Modex	Aircraft Type	Air Wing/ Tail Code	Area of Operations	Operations/ Exercises
Feb–Aug 60	VF-91	100	F8U-2	CVG-9 NG	WestPac	
	VF-92	200	F3H-2			
	VA-93	300	A4D-2			
	VA-94	400	A4D-2			
	VA-95	500	AD-7			
	VAH-6	x	A3D-2			
	VAW-13 Det.M	VR7xx	AD-5Q/AD-5W			
	VCP-61 Det.M	SS9xx	F8U-1P			
	HU-1 Det.M	UPxx	HUP-2			
Jun–Jul 61	VF-91	100	F8U-2	CVG-9 NG	EastPac	
	VF-92	200	F3H-2			
	VA-93	300	A4D-2N			
	VA-94	400	A4D-2N			
	VA-95	500	AD-7			
	VAH-6	x	A3D-2			
	VAW-11 Det.M	RR7xx	AD-5Q/WF-2			
	VCP-63 Det.M	PP9xx	F8U-1P			
	HU-1 Det.M	UPxx	HUP-3			
Aug 61–Mar 62	VF-91	100	F8U-2	CVG-9 NG	WestPac	
	VF-92	200	F3H-2			
	VA-93	300	A4D-2N			
	VA-94	400	A4D-2N			
	VA-95	500	AD-7			
	VAH-6	x	A3D-2			
	VAW-13 Det.M	VR71x	AD-5Q			
	VAW-11 Det.M	RR78x	WF-2			
	VFP-63 Det.M	PP9xx	F8U-1P			
	HU-1 Det.M	UPxx	HUP-3			
Sep–Oct 62	VF-91	100	F-8C	CVG-9 NG	EastPac	
	VA-93	300	A-4C			
	VA-94	400	A-4C			
	VA-95	500	A-1H/J			
	VF-96	600	F-4B			
	VAH-6	x	A-3B			
	VAW-11 Det.M	RR7xx	E-1B			
	VFP-63 Det.M	PP9xx	RF-8A			
	HU-1 Det.M	UPxx	UH-25B			
Nov 62–Jun 63	VF-91	100	F-8C	CVG-9 NG	WestPac	
	VA-93	300	A-4C			
	VA-94	400	A-4C			
	VA-95	500	A-1H/J			
	VF-96	600	F-4B			
	VAH-6	x	A-3B			
	VAW-11 Det.M	RR7xx	E-1B			
	VFP-63 Det.M	PP91x	RF-8A			
	HU-1 Det.M	UPxx	UH-25B			

CV-61 *Ranger* (continued)

Deployment Dates	Squadrons	Modex	Aircraft Type	Air Wing/ Tail Code	Area of Operations	Operations/ Exercises
Jun–Jul 64	VF-92	200	F-4B	CVW-9 NG	HI	
	VA-93	300	A-4C			
	VA-94	400	A-4C			
	VF-96	600	F-4B			
	VAH-6	x	A-3B			
	VAW-11 Det.M	RR7xx	E-1B			
	VFP-63 Det.M	PP9xx	RF-8A			
	HU-1 Det.M	UPxx	UH-2A			
Aug 64–May 65	RVAH-5	100	RA-5C	CVW-9 NG	WestPac Vietnam	Candid Camera
	VF-92	200	F-4B			
	VA-93	300	A-4C			
	VA-94	400	A-4C			
	VA-95	500	A-1H/J			
	VF-96	600	F-4B			
	VAW-13 Det.M	VR710	EA-1F			
	VAW-11 Det.M	RR740	E-1B			
	VAH-2 Det.M	NL810	A-3B			
	HU-1 Det.M	UPxx	UH-2A			
Dec 65–Aug 66	VF-142	200	F-4B	CVW-14 NK	WestPac Vietnam	
	VF-143	300	F-4B			
	VA-55	400	A-4E			
	VA-145	500	A-1H/J			
	VA-146	600	A-4C			
	VAW-11 Det.F	RR73x	E-2A			
	VAH-2 Det.F	ZA810	A-3B			
	RVAH-9	AC900	RA-5C			
	HC-1 Det.F.	UP1-3	UH-2A/B			
Sep–Oct 67	VF-21	100	F-4B	CVW-2 NE	EastPac	
	VA-22	200	A-4C			
	VA-147	300	A-7A			
	VF-154	400	F-4B			
	VA-165	500	A-6A			
	VAH-2 Det.61	ZA600	KA-3B			
	RVAH-6	700	RA-5C			
	VAW-115	750	E-2A			
	HC-1 Det.61	UPxx	UH-2C			
Nov 67–May 68	VF-21	00, 100	F-4B	CVW-2 NE	WestPac Vietnam	
	VA-22	200	A-4C			
	VA-147	300	A-7A			
	VF-154	400	F-4B			
	VA-165	500	A-6A			
	VAH-2 Det.61	ZA600	KA-3B			
	RVAH-6	700	RA-5C			
	VAW-13 Det.61	VR721-723	EKA-3B			
	VAW-115	750	E-2A			
	HC-1 Det.61	UPxx	UH-2C			

Deployment Dates	Squadrons	Modex	Aircraft Type	Air Wing/ Tail Code	Area of Operations	Operations/ Exercises
Oct 68–May 69	VF-21	100	F-4J	CVW-2 NE	WestPac Vietnam	
	VF-154	200	F-4J			
	VA-147	300	A-7A			
	VA-155	400	A-4F			
	VA-165	500	A-6A			
	RVAH-9	600	RA-5C			
	VAH-10 Det.61	610-611	KA-3B			
	VAQ-130 Det.61	614-616	EKA-3B			
	VAW-115	10	E-2A			
	HC-1 Det.61	UP004-006	UH-2C			
	HS-2 Det.	NV300	SH-3A			
Dec 69–Jun 70	VF-21	100	F-4J	CVW-2 NE	WestPac Vietnam	
	VF-154	200	F-4J			
	VA-93	300	A-7B			
	VA-56	400	A-7B			
	VA-196	500	A-6A			
	RVAH-5	600	RA-5C			
	VAQ-134	610-611	KA-3B			
	VAW-115	614-616	EKA-3B			
	HC-1 Det.1	10	E-2A			
		004-006	SH-3A			
Sep 70–Jun 71	VF-21	100	F-4J	CVW-2 NE	WestPac Vietnam	
	VF-154	200	F-4J			
	VA-113	300	A-7E			
	VA-25	400	A-7E			
	VA-145	500	A-6A/A-6C			
	RVAH-1	600	RA-5C			
	VAQ-134	610-616	KA-3B/EKA-3B			
	VAW-111 Det.61	10	E-1B			
	HC-1 Det.1	004-006	SH-3G			
Nov 72–Jun 73	VF-154	100	F-4J	CVW-2 NE	WestPac Vietnam	Linebacker II
	VF-21	200	F-4J			
	VA-113	300	A-7E			
	VA-25	400	A-7E			
	VA-145	500	A-6A/B/C			
	RVAH-5	600	RA-5C			
	VAQ-130 Det.4	610-612	EKA-3B			
	VAW-111 Det.1	014-017	E-1B			
	HC-1 Det.4	004-006	SH-3G			
May–Oct 74	VF-154	100	F-4J	CVW-2 NE	WestPac	
	VF-21	200	F-4J			
	VA-113	300	A-7E			
	VA-25	400	A-7E			
	VA-145	500	A-6A/KA-6D			
	RVAH-13	600	RA-5C			
	VAQ-130 Det.4	610	EKA-3B			
	VQ-1Det.61	PR701,703	EA-3B			
	VAW-112	10	E-2B			
	HC-1 Det.4	xxx	SH-3G			

CV-61 *Ranger* (continued)

Deployment Dates	Squadrons	Modex	Aircraft Type	Air Wing/ Tail Code	Area of Operations	Operations/ Exercises
Apr–May 75	VA-113	300	A-7E	CVW-2 NE	EastPac	
	VA-25	400	A-7E			
	VA-145	500	A-6A/KA-6D			
	HS-4	xxx	SH-3G/D			
Jan–Sep 76	VF-154	100	F-4J	CVW-2 NE	WestPac IO	
	VF-21	200	F-4J			
	VA-113	300	A-7E			
	VA-25	400	A-7E			
	VA-145	500	A-6A/KA-6D			
	RVAH-5	600	RA-5C			
	VAQ-135	630	EA-6B			
	VAW-112	xxx	E-2B			
	HS-4	xxx	SH-3D			
Jan 79	VF-154	100	F-4J	CVW-2 NE	EastPac	
	VF-21	200	F-4J			
	VA-113	300	A-7E			
	VA-25	400	A-7E			
	VA-145	500	A-6E/KA-6D			
	VAW-117	600	E-2B			
	RVAH-7	610	RA-5C			
	VAQ-137	620	EA-6B			
	VS-29	700	S-3A			
	HS-4	730	SH-3H			
Feb–Sep 79	VF-154	100	F-4J	CVW-2 NE	WestPac	
	VF-21	200	F-4J			
	VA-113	300	A-7E			
	VA-25	400	A-7E			
	VA-145	500	A-6E/KA-6D			
	VAW-117	600	E-2B			
	RVAH-7	610	RA-5C			
	VAQ-137	620	EA-6B			
	VS-29	700	S-3A			
	HS-4	730	SH-3H			
Sep 80–May 81	VF-1	100	F-14A	CVW-2 NE	WestPac	
	VF-2	200	F-14A			
	VA-113	300	A-7E			
	VA-25	400	A-7E			
	VA-145	500	A-6E/KA-6D			
	VAW-117	600	E-2B			
	VAQ-137	620	EA-6B			
	VS-37	700	S-3A			
	HS-2	720	SH-3H			

Deployment Dates	Squadrons	Modex	Aircraft Type	Air Wing/ Tail Code	Area of Operations	Operations/ Exercises
Jan–Feb 82	VF-1	100	F-14A	CVW-2 NE	HI	
	VF-2	200	F-14A			
	VA-113	300	A-7E			
	VA-25	400	A-7E			
	VA-145	500	A-6E/KA-6D			
	VAW-116	600-603	E-2C			
	VAQ-137	604-607	EA-6B			
	HS-2	610	SH-3H			
	VS-21	700	S-3A			
Apr–Oct 82	VF-1	100	F-14A	CVW-2 NE	WestPac IO	
	VF-2	200	F-14A			
	VA-113	300	A-7E			
	VA-25	400	A-7E			
	VA-145	500	A-6E/KA-6D			
	VAW-116	600-603	E-2C			
	VAQ-137	604-607	EA-6B			
	HS-2	610	SH-3H			
	VS-21	700	S-3A			
	VQ-1 Det.	PRxxx	EA-3B			
	VRC-50 Det.	RG42x/71x	C-2A/US-3A			
Jul 83–Feb 84	VF-211	100	F-14A	CVW-9 NG	WestPac IO	
	VF-24	200	F-14A			
	VA-192	300	A-7E			
	VA-195	400	A-7E			
	VA-165	500	A-6E/KA-6D			
	VAW-112	600-603	E-2C			
	VAQ-138	604-607	EA-6B			
	HS-8	610	SH-3H			
	VS-33	700	S-3A			
May–Jul 86	VF-1	100	F-14A	CVW-2 NE		RimPac '86
	VF-2	200	F-14A			
	VMA(AW)-121	400	A-6E			
	VA-145	500	A-6E/KA-6D			
	VAW-116	600-603	E-2C			
	VAQ-131	604-607	EA-6B			
	HS-14	610	SH-3H			
	VS-38	700	S-3A			
Aug–Oct 86	VF-1	100	F-14A	CVW-2 NE	NorPac	
	VF-2	200	F-14A			
	VMA(AW)-121	400	A-6E			
	VA-145	500	A-6E/KA-6D			
	VAW-116	600-603	E-2C			
	VAQ-131	604-607	EA-6B			
	HS-14	610	SH-3H			
	VS-38	700	S-3A			

CV-61 *Ranger* (continued)

Deployment Dates	Squadrons	Modex	Aircraft Type	Air Wing/ Tail Code	Area of Operations	Operations/ Exercises
Mar–Apr 87	VF-1	100	F-14A	CVW-2 NE	NorPac	
	VF-2	200	F-14A			
	VMA(AW)-121	400	A-6E			
	VA-145	500	A-6E/KA-6D			
	VAW-116	600-603	E-2C			
	VAQ-131	604-607	EA-6B			
	HS-14	610	SH-3H			
	VS-38	700	S-3A			
Jul–Dec 87	VF-1	100	F-14A	CVW-2 NE	WestPac IO	Earnest Will Nimble Archer
	VF-2	200	F-14A			
	VMA(AW)-121	400	A-6E			
	VA-145	500	A-6E/KA-6D			
	VAW-116	600-603	E-2C			
	VAQ-131	604-607	EA-6B			
	HS-14	610	SH-3H			
	VS-38	700	S-3A			
Feb–Aug 89	VF-1	100	F-14A	CVW-2 NE	WestPac IO	
	VF-2	200	F-14A			
	VMA(AW)-121	400	A-6E			
	VA-145	500	A-6E			
	VAW-116	600-603	E-2C			
	VAQ-131	604-607	EA-6B			
	HS-14	610	SH-3H			
	VS-38	700	S-3A			
	VRC-50Det.	RG71x	US-3A			
Dec 90–Jun 91	VF-1	100	F-14A	CVW-2 NE	WestPac IO Gulf	Desert Shield Desert Storm
	VF-2	200	F-14A			
	VA-155	400	A-6E			
	VA-145	500	A-6E			
	VAW-116	600-603	E-2C			
	VAQ-131	604-607	EA-6B			
	HS-14	610	SH-3H			
	VS-38	700	S-3A			
	VRC-50 Det.	RG4xx	C-2A			
Aug 92–Jan 93	VF-1	100	F-14A	CVW-2 NE	WestPac IO Gulf	Southern Watch Restore Hope
	VF-2	200	F-14A			
	VA-155	400	A-6E			
	VA-145	500	A-6E			
	VAW-116	600-603	E-2C			
	VAQ-131	604-607	EA-6B			
	HS-14	610	SH-3H			
	VS-38	700	S-3A			

CV-62 *Independence*

Deployment Dates	Squadrons	Modex	Aircraft Type	Air Wing/ Tail Code	Area of Operations	Operations/ Exercises
Apr–Jun 59	VF-41	100	F3H-2	CVG-7 AG	Carib	Shakedown
	VF-11	AB200	F8U-1			
	VA-86	400	A4D-2			
	VA-75	500	AD-6			
	VAH-1	600	A3D-2			
	VFP-62 Det.41	91x	F8U-1P			
	HU-2 Det.41	HUxx	HUP-2			
Jun–Jul 60	VF-41	100	F3H-2	CVG-7 AG	Virginia Capes	
	VF-84	200	F8U-2			
	VA-72	300	A4D-2			
	VA-86	400	A4D-2			
	VA-75	500	AD-6			
	VAH-1	600	A3D-2			
	VAW-12 Det.41	7xx	WF-2			
	VAW-33 Det.41	8xx	AD-5Q			
	VFP-62 Det.41	9xx	F8U-1P			
	HU-2 Det.41	HUxx	HUP-2			
Aug 60–Mar 61	VF-41	100	F3H-2	CVG-7 AG	Med	
	VF-84	200	F8U-2			
	VA-72	300	A4D-2			
	VA-86	400	A4D-2			
	VA-75	500	AD-6			
	VAH-1	600	A3D-2			
	VAW-12 Det.41	7xx	WF-2			
	VAW-33 Det.41	8xx	AD-5Q			
	VFP-62 Det.41	91x	F8U-1P			
	HU-2 Det.41	HUxx	HUP-3			
	VMA-224	WKxx	A4D-2			
Jun–Jul 61	VF-41	100	F3H-2	CVG-7 AG	Carib	
	VF-84	200	F8U-2			
	VA-72	300	A4D-2N			
	VA-86	400	A4D-2			
	VA-75	500	AD-6			
	VAH-1	600	A3D-2			
	VAW-12 Det.41	GE7xx	WF-2			
	VAW-33 Det.41	GD8xx	AD-5Q			
	VFP-62 Det.41	GA9xx	F8U-1P			
	HU-2 Det.41	HUxx	HUP-3			
Aug–Dec 61	VF-41	100	F3H-2	CVG-7 AG	Med	
	VF-84	200	F8U-2			
	VA-72	300	A4D-2N			
	VA-86	400	A4D-2			
	VA-75	500	AD-6			
	VAH-1	600	A3D-2			
	VAW-12 Det.41	73x	WF-2			
	VAW-33 Det.41	8xx	AD-5Q			
	VFP-62 Det.41	9xx	F8U-1P			
	HU-2 Det.41	HUxx	HUP-3			

CV-62 *Independence* (continued)

Deployment Dates	Squadrons	Modex	Aircraft Type	Air Wing/ Tail Code	Area of Operations	Operations/ Exercises
Apr–Aug 62	VMF(AW)-115	100	F4D-1	CVG-7 AG	Med	
	VF-84	200	F8U-2			
	VA-72	300	A4D-2N			
	VA-86	400	A4D-2			
	VA-75	500	AD-6			
	VAH-1	600	A3D-2			
	VAW-12 Det.41/62	760	WF-2			
	VAW-33 Det.41/62	8xx	AD-5Q			
	VFP-62 Det.41/62	900	F8U-1P			
	HU-2 Det.41/62	HUxx	HUP-3			
Oct–Nov 62	VF-13	AK100	F-3B	CVG-7 AG	Carib	Cuban Missile Crisis (VA-34 from CVAN-65 & VA-64 to CVAN-65 Nov)
	VF-84	200	F-8C			
	VA-72	300	A-4C			
	VA-34	AC300	A-4C			
	VA-75	500	A-1H			
	VA-64	AF500	A-4C			
	VAH-11 Det.8	AB6xx	A-3B			
	VAW-12 Det.62	7xx	E-1B			
	VFP-62 Det.62	9xx	RF-8A			
	HU-2 Det.62	HUxx	UH-25C			
Aug 63–Mar 64	VF-41	100	F-4B	CVG-7 AG	Med	
	VF-84	200	F-8C			
	VA-72	300	A-4C			
	VA-86	400	A-4C			
	VMA-324	500	A-4B			
	VFP-62 Det.62	600	RF-8A			
	VAH-1	700	A-5A			
	VAW-12 Det.62	710	E-1B			
	VAW-33 Det.62	8xx	EA-1F			
	HU-2 Det.62	HUxx	UH-2A			
Jun–Jul 64	VF-41	100	F-4B	CVW-7	WestLant	
	VA-72	300	A-4E			
	VA-86	400	A-4E			
	VA-75	500	A-6A			
	VAH-11 Det.8	AB600	A-3B			
	VAW-12 Det.62	7xx	E-1B			
	VAW-33 Det.62	8xx	EA-1F			
	VFP-62 Det.62	9xx	RF-8A			
	HU-2 Det.62	HUxx	UH-2A			

Deployment Dates	Squadrons	Modex	Aircraft Type	Air Wing/ Tail Code	Area of Operations	Operations/ Exercises
Sep–Nov 64	VF-41	100	F-4B	CVW-7 AG	NorLant Med	Teamwork
	VA-72	300	A-4E			
	VA-86	400	A-4E			
	VAH-11 Det.8	AB600	A-3B			
	VAW-12 Det.62	7xx	E-1B			
	VAW-33 Det.62	8xx	EA-1F			
	VFP-62 Det.62	9xx	RF-8A			
	HU-2 Det.62	HUxx	UH-2A			
May–Dec 65	VF-41	100	F-4B	CVW-7 AG	WestPac Vietnam	
	VF-84	200	F-4B			
	VA-72	300	A-4E			
	VA-86	400	A-4E			
	VA-75	500	A-6A			
	RVAH-1	600	RA-5C			
	VAW-12 Det.62	710	E-1B			
	VAH-4 Det.62	810	A-3B			
	VAW-13 Det.1	VR7xx	EA-1F			
	VQ-1 Det.	PRxx	EA-3B			
	HU-2 Det.62	HUxx	UH-2A			
Jun 66–Feb 67	VF-41	100	F-4B	CVW-7 AG	Med	
	VF-84	200	F-4B			
	VMA-324	300	A-4E			
	VA-86	400	A-4E			
	VA-75	500	A-6A			
	RVAH-1	600	RA-5C			
	VAW-12 Det.62	7xx	E-1B			
	VAW-33 Det.62	8xx	EA-1F			
	HC-2 Det.62	HUxx	UH-2A			
	VQ-2 Det.	JQxx	EA-3B			
Apr 68–Jan 69	VF-41	100	F-4J	CVW-7 AG	Med	
	VF-84	200	F-4J			
	VA-76	300	A-4C			
	RVAH-7	400	RA-5C			
	VAH-10 Det.62	410	KA-3B			
	VSF-1	500	A-4C			
	VA-64	600	A-4C			
	VAQ-33 Det.62	700	EA-1F			
	VAW-124	740	E-2A			
	HC-2 Det.62	HUxx	UH-2A/B			

CV-62 *Independence* (continued)

Deployment Dates	Squadrons	Modex	Aircraft Type	Air Wing/ Tail Code	Area of Operations	Operations/ Exercises
Sep–Oct 69	VF-102	100	F-4J	CVW-7 AG	NorLant	
	VF-33	200	F-4J			
	VA-106	300	A-4C			
	VAH-10 Det.62	400	KA-3B			
	VSF-1	500	A-4C			
	VA-64	600	A-4C			
	VAW-122	720	E-2A			
	RVAH-12	x	RA-5C			
	HC-2 Det.62	HUxx	UH-2B			
	VRC-40 Det.62	xxx	C-1A			
Jun 70–Feb 71	VF-102	100	F-4J	CVW-7 AG	Med	(VAH-10 redes. VAQ-129 Sep)
	VF-33	200	F-4J			
	VMA-331	300	A-4E			
	VA-65	400	A-6A			
	RVAH-11	600	RA-5C			
	VAH-10 Det.62	610	KA-3B			
	VAW-122	720	E-2A			
	HC-2 Det.62	550	HH-2D			
Sep 71–Mar 72	VF-102	100	F-4J	CVW-7 AG	NorLant Med	
	VF-33	200	F-4J			
	VA-66	300	A-7E			
	VA-65	400	A-6A/KA-6D			
	VA-12	500	A-7E			
	RVAH-12	600	RA-5C			
	VAW-122	720	E-2B			
	HS-6	xxx	SH-3A			
	VRC-40 Det.10	xxx	C-1A			
Jun 73–Jan 74	VF-102	100	F-4J	CVW-7 AG	Med	
	VF-33	200	F-4J			
	VA-66	300	A-7E			
	VA-12	400	A-7E			
	VA-65	500	A-6E/KA-6D			
	RVAH-14	600	RA-5C			
	VAW-122	720	E-2B			
	VS-28	xxx	S-2E			
	HS-5	xxx	SH-3D			
Jul 74–Jan 75	VF-102	100	F-4J	CVW-7 AG	Med	
	VF-33	200	F-4J			
	VA-66	300	A-7E			
	VA-12	400	A-7E			
	VA-65	500	A-6E/KA-6D			
	RVAH-9	600	RA-5C			
	VAQ-132	700	EA-6B			
	VAW-122	720	E-2B			
	VS-31	920	S-2G			
	HS-5	xxx	SH-3D			

Deployment Dates	Squadrons	Modex	Aircraft Type	Air Wing/ Tail Code	Area of Operations	Operations/ Exercises
Oct 75–May 76	VF-102	100	F-4J	CVW-7 AG	NorLant Med	
	VF-33	200	F-4J			
	VA-66	300	A-7E			
	VA-12	400	A-7E			
	VA-65	500	A-6E/KA-6D			
	RVAH-13	600	RA-5C			
	VAQ-132	710	EA-6B			
	VAW-117	770	E-2B			
	HS-5	800	SH-3D			
Mar–Oct 77	VF-102	100	F-4J	CVW-7 AG	Med	
	VF-33	200	F-4J			
	VA-66	300	A-7E			
	VA-12	400	A-7E			
	VA-65	500	A-6E/KA-6D			
	VAW-117	600	E-2B			
	RVAH-12	611-613	RA-5C			
	VAQ-136	614-617	EA-6B			
	VS-31	700	S-3A			
	HS-5	xxx	SH-3D			
Jun–Dec 79	VF-102	100	F-4J	CVW-6 AE	Med	
	VF-33	200	F-4J			
	VA-15	300	A-7E			
	VA-87	400	A-7E			
	VA-176	500	A-6E/KA-6D			
	VS-28	600	S-3A			
	VAQ-130	700	EA-6B			
	VFP-63 Det.	xxx	RF-8G			
	VAW-122	720	E-2C			
	HS-15	xxx	SH-3H			
Nov 80–Jun 81	VF-102	100	F-4J	CVW-6 AE	SoLant IO Med	
	VF-33	200	F-4J			
	VA-15	300	A-7E			
	VA-87	400	A-7E			
	VA-176	500	A-6E/KA-6D			
	VAW-122	600-603	E-2C			
	VAQ-131	604-607	EA-6B			
	HS-15	610-615	SH-3H			
	VFP-63 Det.4	616-620	RF-8G			
	VS-28.	700	S-3A			
Jun–Dec 82	VF-14	100	F-14A	CVW-6 AE	Med	
	VF-32	200	F-14A			
	VA-15	300	A-7E			
	VA-87	400	A-7E			
	VA-176	500	A-6E/KA-6D			
	VAW-122	600-603	E-2C			
	VAQ-131	604-607	EA-6B			
	HS-15	610	SH-3H			
	VS-28.	700	S-3A			

CV-62 *Independence* (continued)

Deployment Dates	Squadrons	Modex	Aircraft Type	Air Wing/ Tail Code	Area of Operations	Operations/ Exercises
Oct 83–Apr 84	VF-14	100	F-14A	CVW-6 AE	Carib Med NorLant	Urgent Fury
	VF-32	200	F-14A			
	VA-15	300	A-7E			
	VA-87	400	A-7E			
	VA-176	500	A-6E/KA-6D			
	VAW-122	600-603	E-2C			
	VAQ-131	604-607	EA-6B			
	HS-15	610	SH-3H			
	VS-28	700	S-3A			
Oct 84–Feb 85	VF-14	100	F-14A	CVW-6 AE	Med IO	
	VF-32	200	F-14A			
	VA-15	300	A-7E			
	VA-87	400	A-7E			
	VA-176	500	A-6E/KA-6D			
	VAW-122	600-603	E-2C			
	VAQ-131	604-607	EA-6B			
	HS-15	610	SH-3H			
	VS-28	700	S-3A			
Aug–Oct 88	VFA-131	AK100	FA-18A	CVW-17 AA	Transit via Cape Horn	(Home port change to San Diego)
	VF-103	200	F-14A			
	VA-155	400	A-6E			
	VA-85	AB500	A-6E/KA-6D			
	VAW-125	600	E-2C			
	HS-3	610	SH-3H			
	VS-30	700	S-3A			
Jun–Dec 90	VF-154	100	F-14A	CVW-14 NK	WestPac IO Persian Gulf	Desert Shield
	VF-21	200	F-14A			
	VFA-113	300	FA-18A			
	VFA-25	400	FA-18A			
	VA-196	500	A-6E			
	VAW-113	600-603	E-2C			
	VAQ-139	604-607	EA-6B			
	HS-8	610	SH-3H			
	VS-37	700	S-3A			
Aug 91	VF-154	100	F-14A	CVW-14 NK	(Transfer to HI)	
	VF-21	200	F-14A			
	VFA-113	300	FA-18C			
	VFA-25	400	FA-18C			
	VA-196	500	A-6E			
	VAW-113	600	E-2C			
	HS-8	610	SH-3H			
	VAQ-139	620	EA-6B			
	VS-21	700	S-3B			
	VRC-30 Det.	RWxx	C-2A			

Deployment Dates	Squadrons	Modex	Aircraft Type	Air Wing/ Tail Code	Area of Operations	Operations/ Exercises
Aug–Sep 91	VF-154	100	F-14A	CVW-5 NF	(Transfer HI to Japan)	
	VF-21	200	F-14A			
	VFA-192	300	FA-18C			
	VFA-195	400	FA-18C			
	VA-115	500	A-6E			
	VAW-115	600	E-2C			
	HS-12	610	SH-3H			
	VAQ-136	620	EA-6B			
	VS-21	700	S-3B			
	VRC-50 Det.	RG4xx/ RG71x	C-2A/US-3A			
Apr–Oct 92	VF-154	100	F-14A	CVW-5 NF	WestPac IO Gulf	Southern Watch
	VF-21	200	F-14A			
	VFA-192	300	FA-18C			
	VFA-195	400	FA-18C			
	VA-115	500	A-6E			
	VAW-115	600	E-2C			
	HS-12	610	SH-3H			
	VAQ-136	620	EA-6B			
	VS-21	700	S-3B			
Feb–Mar 93	VF-154	100	F-14A	CVW-5 NF	WestPac	Team Spirit
	VF-21	200	F-14A			
	VFA-192	300	FA-18C			
	VFA-195	400	FA-18C			
	VA-115	500	A-6E			
	VAW-115	600	E-2C			
	HS-12	610	SH-3H			
	VAQ-136	620	EA-6B			
	VS-21	700	S-3B			
May–Jul 93	VF-154	100	F-14A	CVW-5 NF	WestPac	
	VF-21	200	F-14A			
	VFA-192	300	FA-18C			
	VFA-195	400	FA-18C			
	VA-115	500	A-6E			
	VAW-115	600	E-2C			
	HS-12	610	SH-3H			
	VAQ-136	620	EA-6B			
	VS-21	700	S-3B			
	VQ-5 Det.A	720	ES-3A			
	VRC-50 Det.	RG430	C-2A			

CV-62 *Independence* (continued)

Deployment Dates	Squadrons	Modex	Aircraft Type	Air Wing/ Tail Code	Area of Operations	Operations/ Exercises
Nov 93–Mar 94	VF-154	100	F-14A	CVW-5 NF	WestPac IO Gulf Somalia	Southern Watch
	VF-21	200	F-14A			
	VFA-192	300	FA-18C			
	VFA-195	400	FA-18C			
	VA-115	500	A-6E			
	VAW-115	600	E-2C			
	HS-12	610	SH-3H			
	VAQ-136	620	EA-6B			
	VS-21	700	S-3B			
	VQ-5 Det.A	720	ES-3A			
	VRC-50 Det.	RG430	C-2A			
Jul–Aug 94	VF-154	100	F-14A	CVW-5 NF	WestPac	
	VF-21	200	F-14A			
	VFA-192	300	FA-18C			
	VFA-195	400	FA-18C			
	VA-115	500	A-6E			
	VAW-115	600	E-2C			
	HS-12	610	SH-3H			
	VAQ-136	620	EA-6B			
	VS-21	700	S-3B			
	VQ-5 Det.A	720	ES-3A			
	VRC-50 Det.	RG430	C-2A			
Aug–Nov 95	VF-154	100	F-14A	CVW-5 NF	WestPac IO Gulf	Southern Watch
	VF-21	200	F-14A			
	VFA-192	300	FA-18C			
	VFA-195	400	FA-18C			
	VA-115	500	A-6E			
	VAW-115	600	E-2C			
	HS-14	610	SH-60F/HH-60H			
	VAQ-136	620	EA-6B			
	VS-21	700	S-3B			
	VQ-5 Det.A	720	ES-3A			
	VRC-30 Det.5	430	C-2A			
Feb–Mar 96	VF-154	100	F-14A	CVW-5 NF	WestPac	
	VFA-192	300	FA-18C			
	VFA-195	400	FA-18C			
	VA-115	500	A-6E			
	VAW-115	600	E-2C			
	HS-14	610	SH-60F/HH-60H			
	VAQ-136	620	EA-6B			
	VS-21	700	S-3B			
	VQ-5 Det.A	720	ES-3A			
	VRC-30 Det.5	430	C-2A			

Deployment Dates	Squadrons	Modex	Aircraft Type	Air Wing/ Tail Code	Area of Operations	Operations/ Exercises
Feb–Jun 97	VF-154	100	F-14A	CVW-5 NF	WestPac	Cobra Gold
	VFA-27	200	FA-18C(N)			
	VFA-192	300	FA-18C			
	VFA-195	400	FA-18C			
	VAW-115	600	E-2C			
	HS-14	610	SH-60F/HH-60H			
	VAQ-136	620	EA-6B			
	VS-21	700	S-3B			
	VQ-5 Det.5	720	ES-3A			
	VRC-30 Det.5	430	C-2A			
Jan–Jun 98	VF-154	100	F-14A	CVW-5 NF	IO Gulf	Southern Watch
	VFA-27	200	FA-18C(N)			
	VFA-192	300	FA-18C			
	VFA-195	400	FA-18C			
	VAW-115	600	E-2C			
	HS-14	610	SH-60F/HH-60H			
	VAQ-136	620	EA-6B			
	VS-21	700	S-3B			
	VQ-5 Det.5	720	ES-3A			
	VRC-30 Det.5	430	C-2A			
Jul–Aug 98	VF-211	100	F-14A	CVW-9 NG	(Transfer)	
	VMFA-314	200	FA-18C(N)			
	VFA-146	300	FA-18C			
	VFA-147	400	FA-18C			
	VAW-112	600	E-2C			
	HS-8	610	SH-60F/HH-60H			
	VAQ-138	620	EA-6B			
	VS-33	700	S-3B			
	VRC-30 Det.4	xx	C-2A			

CV-63 *Kitty Hawk*

Deployment Dates	Squadrons	Modex	Aircraft Type	Air Wing/ Tail Code	Area of Operations	Operations/ Exercises
Aug–Nov 61	VF-142	NK200	F8U-1	CVG-11 NH	Carib SoLant EastPac	
	VA-113	300	A4D-2N			
	VA-115	500	AD-6			
	VAH-13	GP600	A3D-2			
	VAW-11 Det.C	RR7xx	WF-2			
	VFP-63 Det.C	PP9xx	F8U-1P			
	HU-1 Det.C	UPxx	HUP-3			
Jun–Jul 62	VF-111	100	F8U-2N	CVG-11 NH	SoCal; HI	
	VA-112	200	A4D-2N			
	VA-113	300	A4D-2N			
	VF-114	400	F4H-1			
	VA-115	500	AD-6/7			
	VAH-13	GP600	A3D-2			
	VAW-11 Det.C	RR7xx	WF-2			
	VFP-63 Det.C	PP9xx	F8U-1P			
	HU-1 Det.C	UPxx	HUP-3			
Sep 62–Apr 63	VF-111	00,100	F8U-2N	CVG-11 NH	WestPac	
	VA-112	200	A4D-2N			
	VA-113	300	A4D-2N			
	VF-114	400	F4H-1			
	VA-115	500	AD-6			
	VAH-13	GP600	A3D-2			
	VAW-11 Det.C	RR7xx	WF-2			
	VFP-63 Det.C	PP9xx	F8U-1P			
	HU-1 Det.C	UPxx	HUP-3			
Aug–Sep 63	VF-111	00,100	F-8D	CVG-11 NH	SoCal HI	
	VA-112	200	A-4C			
	VA-113	300	A-4C			
	VF-114	400	F-4B			
	VA-115	500	A-1H			
	VAH-13	610	A-3B			
	VAW-11 Det.C	RR7xx	E-1B			
	VFP-63 Det.C	PP9xx	RF-8A			
	HU-1 Det.C	UPxx	UH-2A			
Oct 63–Jul 64	VF-111	00,100	F-8D	CVG-11 NH	WestPac Vietnam	Big Dipper Back Pack
	VA-112	200	A-4C			
	VA-113	300	A-4C			
	VF-114	400	F-4B			
	VA-115	500	A-1H/J			
	VAH-13	610	A-3B			
	VAW-11 Det.C	RR7xx	E-1B			
	VFP-63 Det.C	PP92x	RF-8A			
	HU-1 Det.C	UPxx	UH-2A			

Deployment Dates	Squadrons	Modex	Aircraft Type	Air Wing/ Tail Code	Area of Operations	Operations/ Exercises
Oct 65–Jun 66	VF-213	100	F-4G/B	CVW-11 NH	WestPac Vietnam	Rolling Thunder Blue Tree Steel Tiger Tiger Hound
	VA-113	300	A-4C			
	VF-114	400	F-4B			
	VA-115	500	A-1H/J			
	RVAH-13	600	RA-5C			
	VAW-11 Det.C	RR704-707	E-2A			
	VA-85	800	A-6A			
	VAH-4 Det.C	ZBx	A-3B			
	HC-1 Det.C	UPxx	UH-2A/B			
Nov 66–Jun 67	VF-213	100	F-4B	CVW-11 NH	WestPac Vietnam	(VAW-11 Det.C est. as VAW-114 Apr 67)
	VF-114	00,200	F-4B			
	VA-144	300	A-4C			
	VA-112	400	A-4C			
	VA-85	500	A-6A			
	RVAH-13	600	RA-5C			
	VAW-11 Det.C	RR704-707	E-2A			
	VAW-114	xxx	KA-3B			
	VAH-4 Det.C	ZB1-5	UH-2B			
	HC-1 Det.C	UPxx				
Nov 67–Jun 68	VF-213	100	F-4B	CVW-11 NH	WestPac Vietnam	
	VF-114	200	F-4B			
	VA-144	300	A-4E			
	VA-112	400	A-4C			
	VA-75	500	A-6A/B			
	RVAH-11	600	RA-5C			
	VAW-114	740	E-2A			
	VAH-4 Det.63	ZB1-6	KA-3B			
	VAW-13 Det.63	VR010	EA-1F			
	HC-1 Det.63	UPxx	UH-2C			
Dec 68–Sep 69	VF-213	100	F-4B	CVW-11 NH	WestPac Vietnam	
	VF-114	200	F-4B			
	VA-37	300	A-7A			
	VA-105	400	A-7A			
	VA-65	500	A-6A/B			
	RVAH-11	600	RA-5C			
	VAQ-131	611-612	KA-3B			
	VAW-114	614-616	EKA-3B			
	HC-1 Det.63	10	E-2A			
		UPxx	UH-2C			
Nov 70–Jun 71	VF-213	100	F-4J	CVW-11 NH	WestPac Vietnam	
	VF-114	200	F-4J			
	VA-192	300	A-7E			
	VA-195	400	A-7E			
	VA-52	500	A-6A/B			
	RVAH-6	600	RA-5C			
	VAQ-133	610-612/614-616	KA-3B/EKA-3B			
	VAW-114	10	E-2A			
	HC-1 Det.2	004-006	UH-2C			

CV-63 *Kitty Hawk* (continued)

Deployment Dates	Squadrons	Modex	Aircraft Type	Air Wing/ Tail Code	Area of Operations	Operations/ Exercises
Feb–Nov 72	VF-213	100	F-4J	CVW-11 NH	WestPac Vietnam	Freedom Porch Freedom Train Linebacker I
	VF-114	200	F-4J			
	VA-192	300	A-7E			
	VA-195	400	A-7E			
	VA-52	500	A-6A/KA-6D			
	RVAH-7	600	RA-5C			
	VAQ-135 Det.1	610	EKA-3B			
	VAW-114	10	E-2B			
	HC-1 Det.1	004-006	SH-3G			
Nov 73–Jul 74	VF-213	100	F-4J	CVW-11 NH	WestPac IO	
	VF-114	200	F-4J			
	VA-192	300	A-7E			
	VA-195	400	A-7E			
	VA-52	500	A-6A/KA-6D			
	RVAH-7	600	RA-5C			
	VAQ-136	610	EA-6B			
	VS-37	700	S-2G			
	VS-38	710	S-2G			
	HS-4	740	SH-3D			
	VAW-114	10	E-2B			
	VQ-1 Det.63	PRxx	EA-3B			
May–Dec 75	VF-114	100	F-4J	CVW-11 NH	WestPac	
	VF-213	200	F-4J			
	VA-192	300	A-7E			
	VA-195	400	A-7E			
	VA-52	500	A-6E/KA-6D			
	VAW-114	601-604	E-2B			
	RVAH-6	610-612	RA-5C			
	VAQ-136	614-617	EA-6B			
	VS-37	700	S-2G			
	VS-38	710	S-2G			
	HS-8	720	SH-3D			
	VQ-1 Det.63	PR004,005	EA-3B			
Oct 77–May 78	VF-114	100	F-14A	CVW-11 NH	WestPac	
	VF-213	200	F-14A			
	VA-192	300	A-7E			
	VA-195	400	A-7E			
	VA-52	500	A-6E/KA-6D			
	VAW-122	604-607	E-2C			
	RVAH-7	610	RA-5C			
	VAQ-131	620	EA-6B			
	VS-33	700	S-3A/US-3A			
	HS-8	720	SH-3D			
	VQ-1 Det.B	PR004,005	EA-3B			

Deployment Dates	Squadrons	Modex	Aircraft Type	Air Wing/ Tail Code	Area of Operations	Operations/ Exercises
May 79–Feb 80	VF-51	100	F-14A	CVW-15 NL	WestPac IO Arabian Sea	
	VF-111	200	F-14A			
	VA-22	300	A-7E			
	VA-94	400	A-7E			
	VA-52	500	A-6E/KA-6D			
	VAW-114	600	E-2C			
	VFP-63 Det.1	610	RF-8G			
	VAQ-135	624-627	EA-6B			
	VS-21	700	S-3A/US-3A			
	HS-8	720	SH-3H			
	VQ-1 Det.	PR001,007	EA-3B			
Apr–Nov 81	VF-51	100	F-14A	CVW-15 NL	WestPac IO	
	VFP-63 Det.1	115-117	RF-8G			
	VF-111	200	F-14A			
	VA-22	300	A-7E			
	VA-94	400	A-7E			
	VA-52	500	A-6E/KA-6D			
	VAW-114	600-603	E-2C			
	VAQ-135	604-607	EA-6B			
	HS-4	610	SH-3H			
	VS-29	700	S-3A			
	VRC-50 Det.	RG425	C-2A			
Jan–Aug 84	VF-1	100	F-14A	CVW-2 NE	WestPac IO Arabian Sea	Team Spirit '84
	VF-2	200	F-14A			
	VA-146	300	A-7E			
	VA-147	400	A-7E			
	VA-145	500	A-6E/KA-6D			
	VAW-116	600-603	E-2C			
	VAQ-130	604-607	EA-6B			
	HS-2	610	SH-3H			
	VS-38	700	S-3A			
	VQ-1Det.	xxx	EA-3B			
	VRC-50 Det.	RG42x/71x	C-2A/US-3A			
Jul–Dec 85	VF-211	100	F-14A	CVW-9 NG	WestPac IO	
	VF-24	200	F-14A			
	VA-146	300	A-7E			
	VA-147	400	A-7E			
	VA-165	500	A-6E/KA-6D			
	VAW-112	600-603	E-2C			
	VAQ-130	604-607	EA-6B			
	HS-2	610	SH-3H			
	VS-33	700	S-3A			

CV-63 *Kitty Hawk* (continued)

Deployment Dates	Squadrons	Modex	Aircraft Type	Air Wing/ Tail Code	Area of Operations	Operations/ Exercises
Jan–Jun 87	VF-211	100	F-14A	CVW-9 NG	World Cruise	
	VF-24	200	F-14A			
	VA-146	300	A-7E			
	VA-147	400	A-7E			
	VA-165	500	A-6E/KA-6D			
	VA-115 Det.	NF500	A-6E/KA-6D			
	VAW-112	600-603	E-2C			
	VAQ-130	604-607	EA-6B			
	HS-2	610	SH-3H			
	VS-33	700	S-3A			
Apr–Dec 91	VF-51	100	F-14A	CVW-15 NL	Pac IO	Cape Horn
	VF-111	200	F-14A			
	VFA-97	300	FA-18A			
	VFA-27	400	FA-18A			
	VA-52	500	A-6E/KA-6D			
	VAW-114	600	E-2C			
	HS-4	610	SH-60F/HH-60H			
	VAQ-134	620	EA-6B			
	VS-37	700	S-3A			
	VRC-50 Det.	RG42x	C-2A			
Nov 92–May 93	VF-51	100	F-14A	CVW-15 NL	WestPac IO Gulf	Restore Hope Southern Watch
	VF-111	200	F-14A			
	VFA-97	300	FA-18A			
	VFA-27	400	FA-18A			
	VA-52	500	A-6E/KA-6D			
	VAW-114	600	E-2C			
	HS-4	610	SH-60F/HH-60H			
	VAQ-134	620	EA-6B			
	VS-37	700	S-3B			
	VRC-50 Det.	RG42x	C-2A			
Jun–Dec 94	VF-51	100	F-14A	CVW-15 NL	WestPac IO	Southern Watch
	VF-111	200	F-14A			
	VFA-97	300	FA-18A			
	VFA-27	400	FA-18A			
	VA-52	500	A-6E			
	VAW-114	600	E-2C			
	HS-4	610	SH-60F/HH-60H			
	VAQ-134	620	EA-6B			
	VS-37	700	S-3B			
	VQ-5 Det.C	720	ES-3A			
	VRC-30 Det.A	RW30,xx	C-2A			

Deployment Dates	Squadrons	Modex	Aircraft Type	Air Wing/ Tail Code	Area of Operations	Operations/ Exercises
Apr–Oct 96	VF-213	100	F-14A	CVW-11 NH	WestPac	Southern Watch
	VFA-97	200	FA-18A			
	VFA-22	300	FA-18C(N)			
	VFA-94	400	FA-18C(N)			
	VAW-117	600	E-2C			
	HS-6	610	SH-60F/HH-60H			
	VAQ-135	620	EA-6B			
	VS-29	700	S-3B			
	VQ-5 Det.B	720	ES-3A			
	VRC-30 Det.2	xx	C-2A			
Oct 96–Apr 97	VF-213	100	F-14A	CVW-11 NH	WestPac IO Gulf	
	VFA-97	200	FA-18A			
	VFA-22	300	FA-18C(N)			
	VFA-94	400	FA-18C(N)			
	VAW-117	600	E-2C			
	HS-6	610	SH-60F/HH-60H			
	VAQ-135	620	EA-6B			
	VS-29	700	S-3B			
	VQ-5 Det.B	720	ES-3A			
	VRC-30 Det.2	xx	C-2A			
Jul 98	VF-211	100	F-14A	CVW-9 NG	(Transfer)	
	VMFA-314	200	FA-18C(N)			
	VFA-146	300	FA-18C(N)			
	VFA-147	400	FA-18C(N)			
	VAW-112	600	E-2C			
	HS-8	610	SH-60F/HH-60H			
	VAQ-138	620	EA-6B			
	VS-33	700	S-3B			
	VRC-30 Det.4	xx	C-2A			
Jul–Aug 98	VF-154	100	F-14A	CVW-5 NF	(HI to Japan)	
	VFA-27	200	FA-18C(N)			
	VFA-192	300	FA-18C(N)			
	VFA-195	400	FA-18C(N)			
	VAW-115	600	E-2C			
	HS-14	610	SH-60F/HH-60H			
	VAQ-136	620	EA-6B			
	VS-21	700	S-3B			
	VQ-5 Det.5	720	ES-3A			
	VRC-30 Det.5	430	C-2A			
Mar–Aug 99	VFA-27	200	FA-18C(N)	CVW-5 NF	IO Pac	Tandem Thrust Southern Watch
	VFA-192	300	FA-18C(N)			
	VFA-195	400	FA-18C(N)			
	VAQ-136	500	EA-6B			
	VAW-115	600	E-2C			
	HS-14	610	SH-60F/HH-60H			
	VS-21	700	S-3B			
	VRC-30 Det.5	430	C-2A			

CV-63 *Kitty Hawk* (continued)

Deployment Dates	Squadrons	Modex	Aircraft Type	Air Wing/ Tail Code	Area of Operations	Operations/ Exercises
Oct–Nov 99	VFA-27	200	FA-18C(N)	CVW-5 NF	WestPac	
	VFA-192	300	FA-18C(N)			
	VFA-195	400	FA-18C(N)			
	VAQ-136	500	EA-6B			
	VAW-115	600	E-2C			
	HS-14	610	SH-60F/HH-60H			
	VS-21	700	S-3B			
	VRC-30 Det.5	430	C-2A			
Apr–Jun 00	VFA-27	200	FA-18C(N)	CVW-5 NF	WestPac	Cobra Gold 2000
	VFA-192	300	FA-18C(N)			
	VFA-195	400	FA-18C(N)			
	VAQ-136	500	EA-6B			
	VAW-115	600	E-2C			
	HS-14	610	SH-60F/HH-60H			
	VS-21	700	S-3B			
	VRC-30 Det.5	430	C-2A			
Sep–Nov 00	VFA-27	200	FA-18C(N)	CVW-5 NF	WestPac	Contingency Operations
	VFA-192	300	FA-18C(N)			
	VFA-195	400	FA-18C(N)			
	VAQ-136	500	EA-6B			
	VAW-115	600	E-2C			
	HS-14	610	SH-60F/HH-60H			
	VS-21	700	S-3B			
	VRC-30 Det.5	430	C-2A			
Mar–Jun 01	VFA-27	200	FA-18C(N)	CVW-5 NF	WestPac	Tandem Thrust
	VFA-192	300	FA-18C(N)			
	VFA-195	400	FA-18C(N)			
	VAQ-136	500	EA-6B			
	VAW-115	600	E-2C			
	HS-14	610	SH-60F/HH-60H			
	VS-21	700	S-3B			
	VRC-30 Det.5	430	C-2A			
Sep–Dec 01	VFA-192 Det.	300	FA-18C(N)	CVW-5 NF	IO Arabian Sea	Enduring Freedom/ Special Operations
	VFA-195 Det.	400	FA-18C(N)			
	HS-14 Det.	610	SH-60F/HH-60H			
	VS-21 Det.	700	S-3B			
	VRC-30 Det.5	430	C-2A			
Apr–Jun 02	VFA-27	200	FA-18C(N)	CVW-5 NF	WestPac	
	VFA-192	300	FA-18C(N)			
	VFA-195	400	FA-18C(N)			
	VAQ-136	500	EA-6B			
	VAW-115	600	E-2C			
	HS-14	610	SH-60F/HH-60H			
	VS-21	700	S-3B			
	VRC-30 Det.5	430	C-2A			

Deployment Dates	Squadrons	Modex	Aircraft Type	Air Wing/ Tail Code	Area of Operations	Operations/ Exercises
Oct–Dec 02	VFA-27	200	FA-18C(N)	CVW-5 NF	WestPac	Keen Sword '03 ANNUALEX 14G
	VFA-192	300	FA-18C(N)			
	VFA-195	400	FA-18C(N)			
	VAQ-136	500	EA-6B			
	VAW-115	600	E-2C			
	HS-14	610	SH-60F/HH-60H			
	VS-21	700	S-3B			
	VRC-30 Det.5	430	C-2A			
Jan–May 03	VFA-27	200	FA-18C(N)	CVW-5 NF	WestPac IO Gulf	Southern Watch Enduring Freedom Iraqi Freedom
	VFA-192	300	FA-18C(N)			
	VFA-195	400	FA-18C(N)			
	VAQ-136	500	EA-6B			
	VAW-115	600	E-2C			
	HS-14	610	SH-60F/HH-60H			
	VS-21	700	S-3B			
	VRC-30 Det.5	430	C-2A			
Nov–Dec 03	VFA-27	200	FA-18C(N)	CVW-5 NF	WestPac	
	VFA-192	300	FA-18C(N)			
	VFA-195	400	FA-18C(N)			
	VAQ-136	500	EA-6B			
	VAW-115	600	E-2C			
	HS-14	610	SH-60F/HH-60H			
	VS-21	700	S-3B			
	VRC-30 Det.5	430	C-2A			
Feb–May 04	VFA-102	100	FA-18F	CVW-5 NF	WestPac	Foal Eagle '04
	VFA-27	200	FA-18C(N)			
	VFA-192	300	FA-18C(N)			
	VFA-195	400	FA-18C(N)			
	VAQ-136	500	EA-6B			
	VAW-115	600	E-2C			
	HS-14	610	SH-60F/HH-60H			
	VS-21	700	S-3B			
	VRC-30 Det.5	430	C-2A			
Jul–Sep 04	VFA-102	100	FA-18F	CVW-5 NF	WestPac	Summer Pulse 2004 JASEX 2004
	VFA-192	300	FA-18C(N)			
	VFA-195	400	FA-18C(N)			
	VAQ-136	500	EA-6B			
	VAW-115	600	E-2C			
	HS-14	610	SH-60F/HH-60H			
	VS-21	700	S-3B			
	HSL-51 Det.3	TA71x	SH-60B			
	VRC-30 Det.5	430	C-2A			

CV-63 *Kitty Hawk* (continued)

Deployment Dates	Squadrons	Modex	Aircraft Type	Air Wing/ Tail Code	Area of Operations	Operations/ Exercises
Feb–Mar 05	VFA-102	100	FA-18F	CVW-5 NF	WestPac SOJ	Foal Eagle 2005
	VFA-27	200	FA-18E			
	VFA-192	300	FA-18C(N)			
	VFA-195	400	FA-18C(N)			
	VAQ-136	500	EA-6B			
	VAW-115	600	E-2C			
	HS-14	610	SH-60F/HH-60H			
	HSL-51 Det.3	TA7xx	SH-60B			
	VRC-30 Det.5	xx	C-2A			
May–Aug 05	VFA-102	100	FA-18F	CVW-5 NF	Pac	Talisman Sabre '05 Orange Crush JASEX '05
	VFA-27	200	FA-18E			
	VFA-192	300	FA-18C(N)			
	VFA-195	400	FA-18C(N)			
	VAQ-136	500	EA-6B			
	VAW-115	600	E-2C			
	HS-14	610	SH-60F/HH-60H			
	HSL-51 Det.3	TA7xx	SH-60B			
	VRC-30 Det.5	xx	C-2A			
Oct–Dec 05	VFA-102	100	FA-18F	CVW-5 NF	WestPac SOJ	ANNUALEX 17G
	VFA-27	200	FA-18E			
	VFA-192	300	FA-18C(N)			
	VFA-195	400	FA-18C(N)			
	VAQ-136	500	EA-6B			
	VAW-115	600	E-2C			
	HS-14	610	SH-60F/HH-60H			
	HSL-51 Det.3	TA7xx	SH-60B			
	VRC-30 Det.5	xx	C-2A			
Jun–Sep 06	VFA-102	100	FA-18F	CVW-5 NF	WestPac	JASEX 2006 Valiant Shield 2006
	VFA-27	200	FA-18E			
	VFA-192	300	FA-18C(N)			
	VFA-195	400	FA-18C(N)			
	VAQ-136	500	EA-6B			
	VAW-115	600	E-2C			
	HS-14	610	SH-60F/HH-60H			
	HSL-51 Det.3	TA7xx	SH-60B			
	VRC-30 Det.5	xx	C-2A			
Oct–Dec 06	VFA-102	100	FA-18F	CVW-5 NF	WestPac	ANNUALEX 18G
	VFA-27	200	FA-18E			
	VFA-192	300	FA-18C(N)			
	VFA-195	400	FA-18C(N)			
	VAQ-136	500	EA-6B			
	VAW-115	600	E-2C			
	HS-14	610	SH-60F/HH-60H			
	HSL-51 Det.3	TA7xx	SH-60B			
	VRC-30 Det.5	xx	C-2A			

Deployment Dates	Squadrons	Modex	Aircraft Type	Air Wing/ Tail Code	Area of Operations	Operations/ Exercises
May–Sep 07	VFA-102	100	FA-18F	CVW-5 NF	WestPac	Talisman Sabre 2007 Valiant Shield 2007 Malabar 07-2
	VFA-27	200	FA-18E			
	VFA-192	300	FA-18C(N)			
	VFA-195	400	FA-18C(N)			
	VAQ-136	500	EA-6B			
	VAW-115	600	E-2C NP			
	HS-14	610	SH-60F/HH-60H			
	HSL-51 Det.3	TA7xx	SH-60B			
	VRC-30 Det.5	xx	C-2A			
Oct–Nov 07	VFA-102	100	FA-18F	CVW-5 NF	WestPac	ANNUALEX 19G
	VFA-27	200	FA-18E			
	VFA-192	300	FA-18C(N)			
	VFA-195	400	FA-18C(N)			
	VAQ-136	500	EA-6B			
	VAW-115	600	E-2C NP			
	HS-14	610	SH-60F/HH-60H			
	HSL-51 Det.3	TA7xx	SH-60B			
	VRC-30 Det.5	xx	C-2A			
Mar–Apr 08	VFA-102	100	FA-18F	CVW-5 NF	WestPac	
	VFA-27	200	FA-18E			
	VFA-192	300	FA-18C(N)			
	VFA-195	400	FA-18C(N)			
	VAQ-136	500	EA-6B			
	VAW-115	600	E-2C NP			
	HS-14	610	SH-60F/HH-60H			
	HSL-51 Det.3	TA7xx	SH-60B			
	VRC-30 Det.5	xx	C-2A			
May–Aug 08	VFA-102	100	FA-18F	CVW-5 NF	Pac	Home port change RIMPAC 2008
	VFA-27	200	FA-18E			
	VFA-192	300	FA-18C(N)			
	VFA-195	400	FA-18C(N)			
	VAQ-136	500	EA-6B			
	VAW-115	600	E-2C NP			
	HS-14	610	SH-60F/HH-60H			
	HSL-51 Det.3	TA7xx	SH-60B			
	VRC-30 Det.5	xx	C-2A			

CV-64 *Constellation*

Deployment Dates	Squadrons	Modex	Aircraft Type	Air Wing/ Tail Code	Area of Operations	Operations/ Exercises
Mar–May 62	VF-131	100	F3H-2	CVG-13 AE	Virginia Capes Carib	Shakedown
	VF-132	200	F8U-2N			
	VA-133	300	A4D-2			
	VA-134	400	A4D-2			
	VA-135	500	AD-6			
	VAH-10	600	A3D-2			
	VAW-12 Det.64	GE7xx	WF-2			
	VFP-62 Det.64	GA9xx	F8U-1P			
	HU-2 Det.64	HUxx	HUP-2			
Jul–Sep 62	VF-51	100	F8U-2NE	CVG-5 NF	SoLant SoCal	(CVG-5 assigned to CVA-16)
	VA-56	400	A4D-2			
	VA-55	500	A4D-2N			
	VAH-10 Det.B	600	A3D-2			
	VFP-63 Det.B	PP9xx	F8U-1P			
	HU-2 Det.64	HUxx	HUP-2			
Nov 62	VF-141	100	F-8E	CVG-14 NK	EastPac	
	VAH-10	200	A-3B			
	VF-143	300	F-4B			
	VA-144	400	A-4C			
	VA-145	500	A-1H/J			
	VA-146	600	A-4C			
	VAW-11 Det.F	RR7xx	E-1B			
	VFP-63 Det.F	PP9xx	RF-8A			
	HU-1 Det.F	UPxx	UH-25B			
Feb–Sep 63	VF-141	100	F-8E	CVG-14 NK	WestPac	
	VAH-10	200	A-3B			
	VF-143	300	F-4B			
	VA-144	400	A-4C			
	VA-145	500	A-1H/J			
	VA-146	600	A-4C			
	VAW-11 Det.F	RR78x	E-1B			
	VFP-63 Det.F	PP91x	RF-8A			
	HU-1 Det.F	UPxx	UH-25B			
May 64–Feb 65	VAH-10	100	A-3B	CVW-14 NK	WestPac Vietnam	
	VF-142	200	F-4B			
	VF-143	300	F-4B			
	VA-144	400	A-4C			
	VA-145	500	A-1H/J			
	VA-146	600	A-4C			
	VAW-11 Det.F	RR78x	E-1B			
	VFP-63 Det.F	PP93x	RF-8A			
	HU-1 Det.F	UPxx	UH-2A			

Deployment Dates	Squadrons	Modex	Aircraft Type	Air Wing/ Tail Code	Area of Operations	Operations/ Exercises
May–Dec 66	VF-151	100	F-4B	CVW-15 NL	WestPac Vietnam	
	VF-161	200	F-4B			
	VA-153	300	A-4C			
	VA-65	400	A-6A			
	VA-155	500	A-4E			
	VAH-8	600	A-3B			
	RVAH-6	700	RA-5C			
	VAW-11 Det.D	750	E-2A			
	HC-1 Det.D	UPxx	UH-2A/B			
Apr–Dec 67	VAH-8	100	KA-3B	CVW-14 NK	WestPac Vietnam	
	RVAH-12	120	RA-5C			
	VF-142	200	F-4B			
	VF-143	300	F-4B			
	VA-196	400	A-6A			
	VA-55	500	A-4C			
	VA-146	600	A-4C			
	VAW-113	750	E-2A			
	HC-1 Det.F/64	UPxx	UH-2A/B			
May 68–Jan 69	VAH-10 Det.64	101-102	KA-3B	CVW-14 NK	WestPac Vietnam	(VAW-13 redes. VAQ-130 Oct)
	VAW-13 Det.64	110-112; 104-106	EKA-3B			
	RVAH-5	120-125; 110-115	RA-5C			
	VF-142	200	F-4B			
	VF-143	300	F-4B			
	VA-196	400	A-6A/B			
	VA-97	500	A-7A			
	VA-27	600	A-7A			
	VAW-113	730-733; 010-013	E-2A			
	HC-1 Det.64	UP84-86	UH-2C			
Aug 69–May 70	VF-143	100	F-4J	CVW-14 NK	WestPac Vietnam	
	VF-142	200	F-4J			
	VA-97	300	A-7A			
	VA-27	400	A-7A			
	VA-85	500	A-6A/B			
	RVAH-7	600	RA-5C			
	VAQ-133	610-611; 614-616	KA-3B/EKA-3B			
	VAW-113	10	E-2A			
	HC-1 Det.64/5	004-006	SH-3A			

CV-64 *Constellation* (continued)

Deployment Dates	Squadrons	Modex	Aircraft Type	Air Wing/ Tail Code	Area of Operations	Operations/ Exercises
Oct 71–Jun 72	VF-96	100	F-4J	CVW-9 NG	WestPac Vietnam	Freedom Train Freedom Porch Pocket Money Linebacker I
	VF-92	200	F-4J			
	VA-146	300	A-7E			
	VA-147	400	A-7E			
	VA-165	500	A-6A/KA-6D			
	RVAH-11	600	RA-5C			
	VAQ-130 Det.1	614-616	EKA-3B			
	VAW-116	10	E-2B			
	HC-1 Det.3	004-007	SH-3G			
Jan–Oct 73	VF-96	100	F-4J	CVW-9 NG	WestPac Vietnam	
	VF-92	200	F-4J			
	VA-146	300	A-7E			
	VA-147	400	A-7E			
	VA-165	500	A-6A/KA-6D			
	RVAH-12	600	RA-5C			
	VAQ-134	610	EA-6B			
	VAW-116	10	E-2B			
	HS-6 Det.1	004-006	SH-3G			
Jun–Dec 74	VF-96	100	F-4J	CVW-9 NG	WestPac Vietnam IO	Midlink 74
	VF-92	200	F-4J			
	VA-146	300	A-7E			
	VA-147	400	A-7E			
	VA-165	500	A-6A/KA-6D			
	RVAH-5	600	RA-5C			
	VAQ-131	610	EA-6B			
	VAW-116	10	E-2B			
	HS-6	1	SH-3A			
	VQ-1 Det.64	PR03,xx	EA-3B			
Apr–Nov 77	VF-211	100	F-14A	CVW-9 NG	WestPac	
	VF-24	200	F-14A			
	VA-146	300	A-7E			
	VA-147	400	A-7E			
	VA-165	500	A-6E/KA-6D			
	VAW-126	600	E-2C			
	VFP-63 Det.1	610	RF-8G			
	VAQ-132	620	EA-6B			
	VS-21	700	S-3A			
	HS-6	720	SH-3A			
	VQ-1 Det.	PR007,010	EA-3B			

Deployment Dates	Squadrons	Modex	Aircraft Type	Air Wing/ Tail Code	Area of Operations	Operations/ Exercises
Sep 78–May 79	VF-211	100	F-14A	CVW-9 NG	WestPac IO	
	VF-24	200	F-14A			
	VA-146	300	A-7E			
	VA-147	400	A-7E			
	VA-165	500	A-6E/KA-6D			
	VAW-126	600	E-2C			
	VFP-63 Det.3	610	RF-8G			
	VAQ-132	620	EA-6B			
	VS-37	700	S-3A			
	HS-6	720	SH-3H			
	COD	737	US-3A			
Feb–Oct 80	VF-211	100	F-14A	CVW-9 NG	WestPac IO	
	VF-24	200	F-14A			
	VA-146	300	A-7E			
	VA-147	400	A-7E			
	VA-165	500	A-6E/KA-6D			
	VAW-116	600	E-2C			
	VFP-63 Det.3	610	RF-8G			
	VS-38	700	S-3A			
	HS-6	720	SH-3H			
Oct 81–May 82	VQ-1 Det.	PR004	EA-3B	CVW-9 NG	WestPac IO	
	VF-211	100	F-14A			
	VF-24	200	F-14A			
	VA-146	300	A-7E			
	VA-147	400	A-7E			
	VA-165	500	A-6E/KA-6D			
	VAW-112	600-603	E-2C			
	VAQ-134	604-607	EA-6B			
	HS-8	610	SH-3H			
	VS-38	700	S-3A			
	VQ-1 Det.B	115-116	EA-3B			
Feb–Aug 85	VF-154	100	F-14A	CVW-14 NK	WestPac IO	
	VF-21	200	F-14A			
	VFA-113	300	FA-18A			
	VFA-25	400	FA-18A			
	VA-196	500	A-6E/KA-6D			
	VAW-113	600-603	E-2C			
	VAQ-139	604-607	EA-6B			
	HS-8	610	SH-3H			
	VS-37	700	S-3A			
	VQ-1 Det.	NF005, xxx	EA-3B			

CV-64 *Constellation* (continued)

Deployment Dates	Squadrons	Modex	Aircraft Type	Air Wing/ Tail Code	Area of Operations	Operations/ Exercises
Sep–Oct 86	VF-154	100	F-14A	CVW-14 NK	NorPac	
	VF-21	200	F-14A			
	VFA-113	300	FA-18A			
	VFA-25	400	FA-18A			
	VA-196	500	A-6E/KA-6D			
	VAW-113	600-603	E-2C			
	VAQ-139	604-607	EA-6B			
	HS-8	610	SH-3H			
	VS-37	700	S-3A			
Apr–Oct 87	VF-154	100	F-14A	CVW-14 NK	WestPac IO	
	VF-21	200	F-14A			
	VFA-113	300	FA-18A			
	VFA-25	400	FA-18A			
	VA-196	500	A-6E/KA-6D			
	VAW-113	600-603	E-2C			
	VAQ-139	604-607	EA-6B			
	HS-8	610	SH-3H			
	VS-37	700	S-3A			
	VQ-1 Det.B	004,xxx	EA-3B			
Dec 88–Jun 89	VF-154	100	F-14A	CVW-14 NK	WestPac IO	
	VF-21	200	F-14A			
	VFA-113	300	FA-18A			
	VFA-25	400	FA-18A			
	VA-196	500	A-6E/KA-6D			
	VAW-113	600-603	E-2C			
	VAQ-139	604-607	EA-6B			
	HS-8	610	SH-3H			
	VS-37	700	S-3A			
Sep–Oct 89	VF-154	100	F-14A	CVW-14 NK	NorPac	
	VF-21	200	F-14A			
	VFA-113	300	FA-18A			
	VFA-25	400	FA-18A			
	VA-196	500	A-6E/KA-6D			
	VAW-113	600-603	E-2C			
	VAQ-139	604-607	EA-6B			
	HS-8	610	SH-3H			
	VS-37	700	S-3A			
Feb–Apr 90	VF-211	100	F-14B	CVW-9 NG	Cape Horn	West to East Coast
	VF-24	200	F-14B			
	VFA-146	300	FA-18C(N)			
	VFA-147	400	FA-18C(N)			
	VA-165	500	A-6E/KA-6D			
	VAW-112	600-603	E-2C			
	VAQ-138	604-607	EA-6B			
	HS-2	610	SH-60F			
	VS-33	700	S-3A			

Deployment Dates	Squadrons	Modex	Aircraft Type	Air Wing/ Tail Code	Area of Operations	Operations/ Exercises
Mar–Apr 93	VF-74	100	F-14B	CVW-17 AA	Carib	Post SLEP shakedown
	VA-35	500	A-6E/KA-6D			
	VAW-125	600	E-2C			
	VAQ-132	620	EA-6B			
	VS-30	700	S-3B			
	VRC-40 Det.	xx	C-2A			
	HC-16 Det.	BFxx	SH-3H			
May–Jul 93	VFA-151	300	FA-18C(N)	CVW-2 NE	Cape Horn	
	VFA-137	400	FA-18C(N)			
	VA-145	500	A-6E			
	VAW-122	600	E-2C			
	HS-14	610	SH-3H			
	VS-38	700	S-3B			
	VRC-30 Det.	RWxx	C-2A			
	HC-11 Det.	VRxx	CH-46D			
May–Jun 94	VF-2	100	F-14D	CVW-2 NE	Pac	
	VMFA-323	200	FA-18C(N)			
	VFA-151	300	FA-18C(N)			
	VFA-137	400	FA-18C(N)			
	VAW-116	600	E-2C			
	HS-2	610	SH-60F/HH-60H			
	VAQ-131	620	EA-6B			
	VS-38	700	S-3B			
Nov 94–May 95	VF-2	100	F-14D	CVW-2 NE	WestPac IO	Beachcrest '95 Southern Watch
	VMFA-323	200	FA-18C(N)			
	VFA-151	300	FA-18C(N)			
	VFA-137	400	FA-18C(N)			
	VAW-116	600	E-2C			
	HS-2	610	SH-60F/HH-60H			
	VAQ-131	620	EA-6B			
	VS-38	700	S-3B			
	VQ-5 Det.D	720	ES-3A			
	VRC-30 Det.2	RW20,36	C-2A			
Oct –Nov 96	VF-2	100	F-14D	CVW-2 NE	NorPac	
	VMFA-323	200	FA-18C(N)			
	VFA-151	300	FA-18C(N)			
	VFA-137	400	FA-18C(N)			
	VAW-116	600	E-2C			
	HS-2	610	SH-60F/HH-60H			
	VAQ-131	620	EA-6B			
	VS-38	700	S-3B			
	VQ-5 Det.C	720	ES-3A			

CV-64 *Constellation* (continued)

Deployment Dates	Squadrons	Modex	Aircraft Type	Air Wing/ Tail Code	Area of Operations	Operations/ Exercises
Apr–Oct 97	VF-2	100	F-14D	CVW-2 NE	WestPac IO Gulf	Southern Watch Arabian Skies Silent Fury
	VMFA-323	200	FA-18C(N)			
	VFA-151	300	FA-18C(N)			
	VFA-137	400	FA-18C(N)			
	VAW-116	600	E-2C			
	HS-2	610	SH-60F/HH-60H			
	VAQ-131	620	EA-6B			
	VS-38	700	S-3B			
	VQ-5 Det.C	720	ES-3A			
	VRC-30 Det.3	36,37	C-2A			
Jun–Dec 99	VF-2	100	F-14D	CVW-2 NE	WestPac IO Gulf	Southern Watch Red Reef
	VMFA-323	200	FA-18C(N)			
	VFA-151	300	FA-18C(N)			
	VFA-137	400	FA-18C(N)			
	VAQ-131	500	EA-6B			
	VAW-116	600	E-2C			
	HS-2	610	SH-60F/HH-60H			
	VS-38	700	S-3B			
	VRC-30 Det.3	32,33	C-2A			
Mar–Sep 01	VF-2	100	F-14D	CVW-2 NE	WestPac IO Gulf	Southern Watch
	VMFA-323	200	FA-18C(N)			
	VFA-151	300	FA-18C(N)			
	VFA-137	400	FA-18C(N)			
	VAQ-131	500	EA-6B			
	VAW-116	600	E-2C			
	HS-2	610	SH-60F/HH-60H			
	VS-38	700	S-3B			
	VRC-30 Det.2	25,32	C-2A			
	HSL-47 Det.4	TY62,71	SH-60B			
Nov 02–Jun 03	VF-2	100	F-14D	CVW-2 NE	WestPac IO Gulf	Enduring Freedom Iraqi Freedom
	VMFA-323	200	FA-18C(N)			
	VFA-151	300	FA-18C(N)			
	VFA-137	400	FA-18C(N)			
	VAQ-131	500	EA-6B			
	VAW-116	600	E-2C			
	HS-2	610	SH-60F/HH-60H			
	VS-38	700	S-3B			
	VRC-30 Det.2	21,22	C-2A			
	HSL-47 Det.4	TY74,75	SH-60B			

CV-66 *America*

Deployment Dates	Squadrons	Modex	Aircraft Type	Air Wing/ Tail Code	Area of Operations	Operations/ Exercises
May–Jul 65	VF-102	100	F-4B	CVW-6 AE	Lant	Shakedown
	VF-33	200	F-4B			
	VA-66	300	A-4C			
	VA-76	500	A-4C			
	VA-64	600	A-4C			
	VAW-12 Det.66	7xx	E-1B			
	VAW-33 Det.66	GD8xx	EA-1F			
	HU-2 Det.66	HUxx	UH-2A/B			
Nov 65–Jul 66	VF-102	100	F-4B	CVW-6 AE	Med	Fairgame IV
	VF-33	200	F-4B			
	VA-66	300	A-4C			
	RVAH-5	400	RA-5C			
	VA-64	600	A-4C			
	VAW-12 Det.66	71x	E-1B			
	VAW-33 Det.66	8xx	EA-1F			
	HC-2 Det.66	HUxx	UH-2A			
Aug–Sep 66	VF-102	100	F-4B	CVW-6 AE	Carib	
	VF-33	200	F-4B			
	VA-66	300	A-4C			
	RVAH-5	400	RA-5C			
	VA-64	600	A-4C			
	VAW-12 Det.66	GE7xx	E-2A			
	VAW-33 Det.66	GD8xx	EA-1F			
	HC-2 Det.66	HUxx	UH-2A/B			
Nov–Dec 66	VF-102	100	F-4B	CVW-6 AE	WestLant	LANTFLEX '66
	VF-33	200	F-4B			
	VA-66	300	A-4C			
	RVAH-5	400	RA-5C			
	VA-64	600	A-4C			
	VAW-12 Det.66	GE7xx	E-2A			
	VAW-33 Det.66	GD8xx	EA-1F			
	HC-2 Det.66	HUxx	UH-2A/B			
Jan–Sep 67	VF-102	100	F-4B	CVW-6 AE	Med	Poker Hand IV Dawn Clear (VAW-12 redes. VAW-122 Apr)
	VF-33	200	F-4B			
	VA-66	300	A-4C			
	RVAH-5	400	RA-5C			
	VA-36	500	A-4C			
	VA-64	600	A-4C			
	VAH-10 Det.66	20	KA-3B			
	VAW-122	720	E-2A			
	VAW-33 Det.66	810	EA-1F			
	HS-9 Det.66	AW5x	SH-3A			
	HC-2 Det.66	HUxx	UH-2A			

CV-66 *America* (continued)

Deployment Dates	Squadrons	Modex	Aircraft Type	Air Wing/ Tail Code	Area of Operations	Operations/ Exercises
Apr–Dec 68	VF-102	100	F-4J	CVW-6 AE	World Cruise Vietnam	New Boy (VAW-13 redes. VAQ-130 Oct)
	VF-33	200	F-4J			
	VA-82	300	A-7A			
	VA-86	400	A-7A			
	VA-85	500	A-6A/B			
	RVAH-13	600	RA-5C			
	VAW-122	720	E-2A			
	VAH-10 Det.66	011-012	KA-3B			
	VAW-13 Det.66	015-017	EKA-3B			
	HC-2 Det.66	HU04-06	UH-2A/B			
Oct–Nov 69	VF-96	100	F-4J	CVW-9 NG	WestLant	
	VF-92	200	F-4J			
	VA-146	300	A-7E			
	VA-147	400	A-7E			
	VA-165	500	A-6A/B/C			
	RVAH-12	600	RA-5C			
	VAQ-132	61x	KA-3B/EKA-3B			
	VAW-124	740	E-2A			
	HC-2 Det.66	HUxx	UH-2C			
Jan–Mar 70	VF-96	100	F-4J	CVW-9 NG	Carib	
	VF-92	200	F-4J			
	VA-146	300	A-7E			
	VA-147	400	A-7E			
	VA-165	500	A-6A/B			
	RVAH-12	600	RA-5C			
	VAQ-132	61x	KA-3B			
	VAW-124	740	EKA-3B			
	HC-2 Det.66	HUxx	E-2A			
			UH-2C			
Apr–Dec 70	VF-96	100	F-4J	CVW-9 NG	WestPac Vietnam	Blue Sky Commando Tiger Autumn Flower
	VF-92	200	F-4J			
	VA-146	300	A-7E			
	VA-147	400	A-7E			
	VA-165	500	A-6A/B/C			
	RVAH-12	600	RA-5C			
	VAQ-132	610-611	KA-3B			
	VAW-124	615-617	EKA-3B			
	HC-2 Det.66	740;010	E-2A			
	VQ-1 Det.	HU004-006	UH-2C			
		PRxx	EA-3B			
Jul–Dec 71	VF-101 Det.66	100	F-4J	CVW-8 AJ	Med	PHIBLEX 2-71 National Week X Deep Furrow 71 National Week XI Ile D'Or
	VMFA-333	200	F-4J			
	VA-82	300	A-7E			
	VA-86	400	A-7E			
	VA-35	500	A-6A/KA-6D			
	RVAH-13	600	RA-5C			
	VAQ-135 Det.2	610	KA-3B/EKA-3B			
	VAW-124	10	E-2A;E-2B			
	HC-2 Det.	HUxx	HH-2D			

Deployment Dates	Squadrons	Modex	Aircraft Type	Air Wing/ Tail Code	Area of Operations	Operations/ Exercises
Jun 72–Mar 73	VF-74	100	F-4J	CVW-8 AJ	WestPac Vietnam	Linebacker I
	VMFA-333	200	F-4J			
	VA-82	300	A-7C			
	VA-86	400	A-7C			
	VA-35	500	A-6A/C/KA-6D			
	RVAH-6	600	RA-5C			
	VAQ-132	610	EA-6B			
	VAW-124	740	E-2B			
	HC-2 Det.66	xxx	SH-3G			
Jan–Aug 74	VF-143	100	F-4J	CVW-8 AJ	Med	National Week XVI PHIBLEX 9-74 Dawn Patrol Shahbaz Flaming Lance
	VF-142	200	F-4J			
	VA-82	300	A-7C			
	VA-86	400	A-7C			
	VA-35	500	A-6E/KA-6D			
	RVAH-1	600	RA-5C			
	VAQ-133	610	EA-6B			
	VAW-124	740	E-2B			
	HC-2 Det.66	00x	SH-3G			
Sep–Oct 74	VF-213	NH100	F-4J	CVW-8 AJ	NorLant	Northern Merger
	VF-103	AC200	F-4J			
	VA-82	300	A-7C			
	VA-86	400	A-7C			
	VA-35	500	A-6E/KA-6D			
	RVAH-1	600	RA-5C			
	VMCJ-2 Det.B	CY610	EA-6A			
	VAW-126	AA760	E-2B			
	HC-2 Det.66	003-005	SH-3G			
Apr–Oct 76	VF-143	100	F-14A	CVW-6 AE	Med	Open Gate Fluid Drive National Week XXI Poop Deck 76 Display Determination
	VF-142	200	F-14A			
	VA-15	300	A-7E			
	VA-87	400	A-7E			
	VA-176	500	A-6E/KA-6D			
	VS-28	600	S-3A			
	VAQ-137	700	EA-6B			
	VFP-63 Det.5	710	RF-8G			
	VAW-124	720	E-2C			
	HS-15	0	SH-3H			
Jun–Jul 77	VF-143	100	F-14A	CVW-6 AE	SoLant	
	VF-142	200	F-14A			
	VA-15	300	A-7E			
	VA-87	400	A-7E			
	VA-176	500	A-6E/KA-6D			
	VS-28	600	S-3A			
	VAQ-137	700	EA-6B			
	VFP-63 Det.5	710	RF-8G			
	VAW-124	720	E-2C			
	HS-15	0	SH-3H			

CV-66 *America* (continued)

Deployment Dates	Squadrons	Modex	Aircraft Type	Air Wing/ Tail Code	Area of Operations	Operations/ Exercises
Sep 77–Apr 78	VF-143	100	F-14A	CVW-6 AE	Med	National Week
	VF-142	200	F-14A			
	VA-15	300	A-7E			
	VA-87	400	A-7E			
	VA-176	500	A-6E/KA-6D			
	VS-28	600	S-3A			
	VAQ-137	700	EA-6B			
	VFP-63 Det.5	710	RF-8G			
	VAW-124	720	E-2C			
	HS-15	0	SH-3H			
Mar–Sep 79	VF-114	100	F-14A	CVW-11 NH	Med	National Week XXVII
	VF-213	200	F-14A			
	VA-192	300	A-7E			
	VA-195	400	A-7E			
	VA-95	500	A-6E/KA-6D			
	VAW-124	600	E-2C			
	VFP-63 Det.4	610	RF-8G			
	VAQ-131	620	EA-6B			
	VS-33	700	S-3A			
	HS-12	740	SH-3H			
Apr–Nov 81	VF-114	100	F-14A	CVW-11 NH	Med IO	Daily Double Weapons Week
	VF-213	200	F-14A			
	VA-192	300	A-7E			
	VA-195	400	A-7E			
	VA-95	500	A-6E/KA-6D			
	VAW-123	600-603	E-2C			
	VAQ-133	604-607	EA-6B			
	HS-12	610	SH-3H			
	VS-33	700	S-3A			
	VQ-2 Det.	JQxx	EA-3B			
	VR-24 Det.	JMxx	C-2A			
May–Jul 82	VF-102	100	F-14A	CVW-1 AB	Lant	
	VF-33	200	F-14A			
	VA-46	300	A-7E			
	VA-72	400	A-7E			
	VA-34	500	A-6E/KA-6D			
	VAW-123	600	E-2C			
	HS-11	610	SH-3H			
	VMAQ-2 Det.Y	620	EA-6B			
	VS-32	700	S-3A			

Deployment Dates	Squadrons	Modex	Aircraft Type	Air Wing/ Tail Code	Area of Operations	Operations/ Exercises
Aug–Nov 82	VF-102	100	F-14A	CVW-1 AB	NorLant Med Carib	United Effort Northern Wedding '82 Display Determination
	VF-33	200	F-14A			
	VA-46	300	A-7E			
	VA-72	400	A-7E			
	VA-34	500	A-6E/KA-6D			
	VAW-123	600-603	E-2C			
	VAQ-135	604-607	EA-6B			
	HS-11	610	SH-3H			
	VS-32	700	S-3A			
Dec 82–Jun 83	VF-102	100	F-14A	CVW-1 AB	Med IO	Beacon Flash 83 Weapons Week Beacon Flash 83-4 Beacon Flash 83-5
	VF-33	200	F-14A			
	VA-46	300	A-7E			
	VA-72	400	A-7E			
	VA-34	500	A-6E/KA-6D			
	VAW-123	600-603	E-2C			
	VAQ-136	604-607	EA-6B			
	HS-11	610	SH-3H			
	VS-32	700	S-3A			
Apr–Nov 84	VF-102	100	F-14A	CVW-1 AB	Carib Med IO	Ocean Venture Display Determination
	VF-33	200	F-14A			
	VA-46	300	A-7E			
	VA-72	400	A-7E			
	VA-34	500	A-6E/KA-6D			
	VAW-123	600-603	E-2C			
	VAQ-135	604-607	EA-6B			
	HS-11	610	SH-3H			
	VS-32	700	S-3A			
	VQ-2 Det.A	JQxx	EA-3B			
	VRC-50 Det.	RG424	C-2A			
Aug–Oct 85	VF-102	100	F-14A	CVW-1 AB	NorLant	
	VF-33	200	F-14A			
	VA-46	300	A-7E			
	VA-72	400	A-7E			
	VA-34	500	A-6E/KA-6D			
	VAW-123	600-603	E-2C			
	VAQ-135	604-607	EA-6B			
	HS-11	610	SH-3H			
	VS-32	700	S-3A			

CV-66 *America* (continued)

Deployment Dates	Squadrons	Modex	Aircraft Type	Air Wing/ Tail Code	Area of Operations	Operations/ Exercises
Mar–Sep 86	VF-102	100	F-14A	CVW-1 AB	Med	Attain Document III Eldorado Canyon Distant Hammer Poop Deck Trident National Week
	VF-33	200	F-14A			
	VA-46	300	A-7E			
	VA-72	400	A-7E			
	VA-34	500	A-6E/KA-6D			
	VAW-123	600-603	E-2C			
	VMAQ-2 Det.Y	604-607	EA-6B			
	HS-11	610	SH-3H			
	VS-32	700	S-3A			
	VQ-2Det.	JQxx	EA-3B			
Mar–May 88	VF-102	100	F-14A	CVW-1 AB	SoLant	
	VF-33	200	F-14A			
	VFA-82	300	FA-18C			
	VFA-86	400	FA-18C			
	VA-85	500	A-6E/KA-6D			
	VAW-123	600-603	E-2C			
	HS-11	610	SH-3H			
	VS-32	700	S-3A			
Feb–Apr 89	VF-102	100	F-14A	CVW-1 AB	Carib NorLant	North Star
	VF-33	200	F-14A			
	VFA-82	300	FA-18C			
	VFA-86	400	FA-18C			
	VA-85	500	A-6E/KA-6D			
	VAW-123	600-603	E-2C			
	VAQ-137	604-607	EA-6B			
	HS-11	610	SH-3H			
	VS-32	700	S-3A			
May–Nov 89	VF-102	100	F-14A	CVW-1 AB	Med IO Gulf	
	VF-33	200	F-14A			
	VFA-82	300	FA-18C			
	VFA-86	400	FA-18C			
	VA-85	500	A-6E/KA-6D			
	VAW-123	600-603	E-2C			
	VAQ-137	604-607	EA-6B			
	HS-11	610	SH-3H			
	VS-32	700	S-3A			
Dec 90–Apr 91	VF-102	100	F-14A	CVW-1 AB	Med Red Sea Gulf	Desert Shield Desert Storm
	VF-33	200	F-14A			
	VFA-82	300	FA-18C			
	VFA-86	400	FA-18C			
	VA-85	500	A-6E/KA-6D			
	VAW-123	600	E-2C			
	HS-11	610	SH-3H			
	VAQ-137	620	EA-6B			
	VS-32	700	S-3B			

Deployment Dates	Squadrons	Modex	Aircraft Type	Air Wing/ Tail Code	Area of Operations	Operations/ Exercises
Aug–Oct 91	VF-102	100	F-14A	CVW-1 AB	Lant	North Star
	VF-33	200	F-14A			
	VFA-82	300	FA-18C			
	VFA-86	400	FA-18C			
	VA-85	500	A-6E/KA-6D			
	VAW-123	600	E-2C			
	HS-11	610	SH-3H			
	VAQ-137	620	EA-6B			
	VS-32	700	S-3B			
Dec 91–Jun 92	VF-102	100	F-14A	CVW-1 AB	Med IO Red Sea Gulf	
	VF-33	200	F-14A			
	VFA-82	300	FA-18C			
	VFA-86	400	FA-18C			
	VA-85	500	A-6E/KA-6D			
	VAW-123	600	E-2C			
	HS-11	610	SH-3H			
	VAQ-137	620	EA-6B			
	VS-32	700	S-3B			
Aug 93–Feb 94	VF-102	100	F-14A	CVW-1 AB	Med Gulf	Deny Flight Provide Promise Sharp Guard Continue Hope Southern Watch
	VFA-82	300	FA-18C			
	VFA-86	400	FA-18C			
	VA-85	500	A-6E			
	VAW-123	600	E-2C			
	HS-11	610	SH-3H			
	VAQ-137	620	EA-6B			
	VS-32	700	S-3B			
	VRC-40 Det.3	45,46	C-2A			
	HMM-162 Det.A	YSxx	CH-46E			
Sep–Oct 94	N/A	N/A	N/A	N/A	Haiti	Uphold Democracy (Army helicopters embarked)
Aug 95–Feb 96	VF-102	100	F-14B	CVW-1 AB	Med Gulf	Deny Flight Deliberate Force Southern Watch Joint Endeavor Decisive Edge
	VMFA-251	200	FA-18C(N)			
	VFA-82	300	FA-18C			
	VFA-86	400	FA-18C			
	VAW-123	600	E-2C			
	HS-11	610	SH-60F/HH-60H			
	VMAQ-3	620	EA-6B			
	VS-32	700	S-3B			
	VQ-6 Det.A	760	ES-3A			
	VRC-40 Det.4	51,52	C-2A			

CV-67 *John F. Kennedy*

Deployment Dates	Squadrons	Modex	Aircraft Type	Air Wing/ Tail Code	Area of Operations	Operations/ Exercises
Nov–Dec 68	VF-14	100	F-4B	CVW-1 AB	Gtmo	Shakedown
	VF-32	200	F-4B			
	VA-83	300	A-4C			
	VA-81	400	A-4C			
	VAW-121 Det.67	GE7xx	E-1B			
	HC-2 Det.67	HUxx	UH-2A/B			
Feb–Mar 69	VF-14	100	F-4B	CVW-1 AB	Virginia Capes	
	VF-32	200	F-4B			
	VA-83	300	A-4C			
	VA-81	400	A-4C			
	VA-95	500	A-4C			
	RVAH-14	600	RA-5C			
	VAH-10 Det.67	650	KA-3B/EKA-3B			
	VAW-121 Det.67	700	E-1B			
	VAQ-33 Det.67	750	EA-1F			
	HC-2 Det.67	10	UH-2C			
Apr–Dec 69	VF-14	100	F-4B	CVW-1 AB	Med	
	VF-32	200	F-4B			
	VA-83	300	A-4C			
	VA-81	400	A-4C			
	VA-95	500	A-4C			
	RVAH-14	600	RA-5C			
	VAH-10 Det.67	650	KA-3B/EKA-3B			
	VAW-121 Det.67	700	E-1B			
	VAQ-33 Det.67	750	EA-1F			
	HC-2 Det.67	10	UH-2C			
Sep 70–Mar 71	VF-14	100	F-4B	CVW-1 AB	Carib NorLant Med	
	VF-32	200	F-4B			
	VA-46	300	A-7B			
	VA-72	400	A-7B			
	VA-34	500	A-6A/B			
	RVAH-14	600	RA-5C			
	VAQ-131	610	KA-3B/EKA-3B			
	VAW-125	750	E-2B			
	HC-2 Det.67	10	HH-2D			
Aug 71	VF-201	100	F-8H	CVWR-20 AF	WestLant	
	VF-202	200	F-8H			
	VA-203	300	A-4L			
	VA-204	400	A-4L			
	VA-205	500	A-4L			
	VFP-206	600	RF-8G			
	VAQ-208	610	KA-3B			
	VAW-207	700	E-1B			
	HC-2 Det.67	AB010	HH-2D			

Deployment Dates	Squadrons	Modex	Aircraft Type	Air Wing/ Tail Code	Area of Operations	Operations/ Exercises
Nov 71–May 72	VF-14	100	F-4B	CVW-1 AB	NorLant Med	Strong Express
	VF-32	200	F-4B			
	VA-46	300	A-7B			
	VA-72	400	A-7B			
	VA-34	500	A-6A/B/C/KA-6D			
	RVAH-14	600	RA-5C			
	VAQ-135 Det.4	610	EKA-3B			
	VAW-125	750	E-2B			
	HC-2 Det.67	10	HH-2D			
Sep–Oct 72	VF-14	100	F-4B	CVW-1 AB		
	VF-32	200	F-4B			
	VA-46	300	A-7B			
	VA-72	400	A-7B			
	VA-34	500	A-6A/B/KA-6D			
	RVAH-14	600	RA-5C			
	VAQ-135 Det.4	610	EKA-3B			
	VAW-125	750	E-2B			
	HC-2 Det.67	10	HH-2D			
Apr–Dec 73	VF-14	100	F-4B	CVW-1 AB	NorLant Med	(VAQ-135 redes. VAQ-130 Aug)
	VF-32	200	F-4B			
	VA-46	300	A-7B			
	VA-72	400	A-7B			
	VA-34	500	A-6A/B/C/KA-6D			
	RVAH-11	600	RA-5C			
	VAQ-135 Det.4	610	EKA-3B			
	VAW-125	750	E-2B			
	HC-2 Det.67	10	SH-3G			
	VRC-40 Det.	xxx	C-1A			
Apr–May 75	VF-14	100	F-14A	CVW-1 AB	WestLant	
	VF-32	200	F-14A			
	VA-46	300	A-7B			
	VA-72	400	A-7B			
	VA-34	500	A-6A/C/KA-6D			
	VS-21	600	S-3A			
	RVAH-1	620	RA-5C			
	VAQ-133	700	EA-6B			
	VAW-125	750	E-2C			
	HS-11	000,010-016	SH-3D			
Jun 75–Jan 76	VF-14	100	F-14A	CVW-1 AB	Med	
	VF-32	200	F-14A			
	VA-46	300	A-7B			
	VA-72	400	A-7B			
	VA-34	500	A-6A/C/KA-6D			
	VS-21	600	S-3A			
	RVAH-1	620	RA-5C			
	VAQ-133	700	EA-6B			
	VAW-125	750	E-2C			
	HS-11	000,010-016	SH-3D			

CV-67 *John F. Kennedy* (continued)

Deployment Dates	Squadrons	Modex	Aircraft Type	Air Wing/ Tail Code	Area of Operations	Operations/ Exercises
Jun–Jul 76	VF-14	100	F-14A	CVW-1 AB	WestLant	
	VF-32	200	F-14A			
	VA-46	300	A-7B			
	VA-72	400	A-7B			
	VA-34	500	A-6E/KA-6D			
	VFP-63 Det.2	600	RF-8G			
	VAQ-133	610	EA-6B			
	VS-32	700	S-3A			
	HS-11	730	SH-3D			
	VAW-125	10	E-2C			
Sep–Nov 76	VF-14	100	F-14A	CVW-1 AB	Med	
	VF-32	200	F-14A			
	VA-46	300	A-7B			
	VA-72	400	A-7B			
	VA-34	500	A-6E/KA-6D			
	VFP-63 Det.2	600	RF-8G			
	VAQ-133	610	EA-6B			
	VS-32	700	S-3A			
	HS-11	730	SH-3D			
	VAW-125	10	E-2C			
Jan–Aug 77	VF-14	100	F-14A	CVW-1 AB	Med	
	VF-32	200	F-14A			
	VA-46	300	A-7B			
	VA-72	400	A-7B			
	VA-34	500	A-6E/KA-6D			
	VFP-63 Det.2	600	RF-8G			
	VAQ-133	610	EA-6B			
	VS-32	700	S-3A			
	HS-11	730	SH-3D			
	VAW-125	10	E-2C			
Jun 78–Feb 79	VF-14	100	F-14A	CVW-1 AB	Med	
	VF-32	200	F-14A			
	VA-46	300	A-7E			
	VA-72	400	A-7E			
	VA-34	500	A-6E/KA-6D			
	VAQ-133	610	EA-6B			
	VFP-63 Det.2	620	RF-8G			
	VS-32	700	S-3A			
	HS-11	730	SH-3D			
	VAW-125	10	E-2C			

Deployment Dates	Squadrons	Modex	Aircraft Type	Air Wing/ Tail Code	Area of Operations	Operations/ Exercises
Aug 80–Mar 81	VF-14	100	F-14A	CVW-1 AB	Med	
	VF-32	200	F-14A			
	VA-46	300	A-7E			
	VA-72	400	A-7E			
	VA-34	500	A-6E/KA-6D			
	VAQ-138	610	EA-6B			
	VS-32	700	S-3A			
	HS-11	730	SH-3H			
	VAW-126	10	E-2C			
	VQ-2 Det.	JQxx	EA-3B			
Oct–Dec 81	VF-11	100	F-14A	CVW-3 AC	Carib	
	VF-31	200	F-14A			
	VA-37	300	A-7E			
	VA-105	400	A-7E			
	VA-75	500	A-6E/KA-6D			
	VAW-126	600-603	E-2C			
	VAQ-138	604-607	EA-6B			
	HS-7	610	SH-3H			
	VS-22	700	S-3A			
Jan–Jul 82	VF-11	100	F-14A	CVW-3 AC	Med IO	
	VF-31	200	F-14A			
	VA-37	300	A-7E			
	VA-105	400	A-7E			
	VA-75	500	A-6E/KA-6D			
	VAW-126	600-603	E-2C			
	VAQ-138	604-607	EA-6B			
	HS-7	610	SH-3H			
	VS-22	700	S-3A			
Apr–Jul 83	VF-11	100	F-14A	CVW-3 AC	NorLant	
	VF-31	200	F-14A			
	VA-75	500	A-6E/KA-6D			
	VA-85	540	A-6E			
	VAW-126	600-603	E-2C			
	VAQ-137	604-607	EA-6B			
	HS-7	610	SH-3H			
	VS-22	700	S-3A			
Sep 83–May 84	VF-11	100	F-14A	CVW-3 AC	SoLant Med	
	VF-31	200	F-14A			
	VA-75	500	A-6E/KA-6D			
	VA-85	540	A-6E			
	VAW-126	600-603	E-2C			
	VAQ-137	604-607	EA-6B			
	HS-7	610	SH-3H			
	VS-22	700	S-3A			

CV-67 *John F. Kennedy* (continued)

Deployment Dates	Squadrons	Modex	Aircraft Type	Air Wing/ Tail Code	Area of Operations	Operations/ Exercises
Aug 86–Mar 87	VF-14	100	F-14A	CVW-3 AC	Med	
	VF-32	200	F-14A			
	VA-66 Det.	300	A-7E			
	VA-75	500	A-6E/KA-6D			
	VMA(AW)-533	540	A-6E			
	VAW-126	600-603	E-2C			
	VAQ-140	604-607	EA-6B			
	HS-7	610	SH-3H			
	VS-22	700	S-3A			
	VQ-2 Det.	JQ14,16	EA-3B			
Aug 88–Feb 89	VF-14	100	F-14A	CVW-3 AC	Med	
	VF-32	200	F-14A			
	VA-75	500	A-6E/KA-6D			
	VMA(AW)-533	540	A-6E			
	VAW-126	600-603	E-2C			
	VAQ-130	604-607	EA-6B			
	HS-7	610	SH-3H			
	VS-22	700	S-3A			
Aug 90–Mar 91	VF-14	100	F-14A	CVW-3 AC	Med Red Sea	Desert Shield Desert Storm
	VF-32	200	F-14A			
	VA-46	300	A-7E			
	VA-72	400	A-7E			
	VA-75	500	A-6E/KA-6D			
	VAW-126	600	E-2C			
	HS-7	610	SH-3H			
	VAQ-130	620	EA-6B			
	VS-22	700	S-3B			
Oct 92–Apr 93	VF-14	100	F-14A	CVW-3 AC	Med	Provide Promise
	VF-32	200	F-14A			
	VFA-37	300	FA-18C(N)			
	VFA-105	400	FA-18C(N)			
	VA-75	500	A-6E			
	VAW-126	600	E-2C			
	HS-7	610	SH-3H			
	VAQ-130	620	EA-6B			
	VS-22	700	S-3B			
	VRC-40 Det.1	45,47	C-2A			
May–Jul 96	VF-41	100	F-14A	CVW-8 AJ	NorLant Carib	
	VF-14	200	F-14A			
	VFA-15	300	FA-18C(N)			
	VFA-87	400	FA-18C(N)			
	VAW-124	600	E-2C			
	HS-3	610	SH-60F/HH-60H			
	VAQ-141	620	EA-6B			
	VS-24	700	S-3B			

Deployment Dates	Squadrons	Modex	Aircraft Type	Air Wing/ Tail Code	Area of Operations	Operations/ Exercises
Apr–Oct 97	VF-41	100	F-14A	CVW-8 AJ	Med Gulf	Deliberate Guard Southern Watch Dynamic Mix Infinite Acclaim Beacon Flash INVITEX '97
	VF-14	200	F-14A			
	VFA-15	300	FA-18C(N)			
	VFA-87	400	FA-18C(N)			
	VAW-124	600	E-2C			
	HS-3	610	SH-60F/HH-60H			
	VAQ-141	620	EA-6B			
	VS-24	700	S-3B			
	VQ-6 Det.A	760	ES-3A			
	VRC-40 Det.4	xx	C-2A			
Sep 99–Mar 00	VF-102	100	F-14B	CVW-1 AB	Med Gulf	Bright Star Southern Watch
	VMFA-251	200	FA-18C(N)			
	VFA-82	300	FA-18C(N)			
	VFA-86	400	FA-18C			
	VAQ-137	500	EA-6B			
	VAW-123	600	E-2C			
	HS-11	610	SH-60F/HH-60H			
	VS-32	700	S-3B			
	VRC-40 Det.2	xx	C-2A			
Feb–Aug 02	VF-143	100	F-14B	CVW-7 AG	Med Gulf	Southern Watch Enduring Freedom
	VF-11	200	F-14B			
	VFA-136	300	FA-18C(N)			
	VFA-131	400	FA-18C(N)			
	VAQ-140	500	EA-6B			
	VAW-121	600	E-2C			
	HS-5	610	SH-60F/HH-60H			
	VS-31	700	S-3B			
	VRC-40 Det.3	41,52	C-2A			
Jun–Dec 04	VF-103	100	F-14B	CVW-17 AA	Med Gulf	Enduring Freedom Iraqi Freedom Summer Pulse 2004
	VFA-34	200	FA-18C(N)			
	VFA-83	300	FA-18C(N)			
	VFA-81	400	FA-18C			
	VAQ-132	500	EA-6B			
	VAW-125	600	E-2C Hawkeye 2000			
	HS-15	610	SH-60F/HH-60H			
	VS-30	700	S-3B			
	VRC-40 Det.4	xx	C-2A			

Notes

Chapter 1. Super Carrier

1. Friedman, Norman, *U.S. Aircraft Carriers*, pp. 255–285.
2. Faltum, Andrew, *The* Essex *Class Aircraft Carriers*, p. 134.
3. Faltum, Andrew, *The* Essex *Class Aircraft Carriers*, p. 133.
4. The concave mirror, source light combination was replaced with a series of Fresnel lenses. The Fresnel Lens Optical Landing System (FLOLS) has not changed much, except for improvements in the inertial stabilization system.
5. The North American AJ Savage (later A-2) was a carrier-based nuclear attack bomber that entered service in 1949. It featured two Pratt & Whitney R-2800 piston engines with an Allison J33 turbojet in the rear fuselage. It was soon replaced in the nuclear strike role by the A3D Skywarrior, but continued to serve in photo reconnaissance and tanker versions.
6. Until 1976, the Fiscal Year began on 1 July of the previous Calendar Year. In 1976 a three-month transition period was added and Fiscal Years now begin on 1 October of the previous Calendar Year.
7. Polmar, Norman, *Aircraft Carriers*, p. 166.
8. Polmar, Norman, *Aircraft Carriers*, p. 99.
9. Friedman, Norman, *U.S. Aircraft Carriers*, p. 256. The carrier force levels have fluctuated over the years for political and budgetary reasons as much as because of operational requirements. The peak level of 15 carriers was in 1991.
10. A ship's displacement, or displacement tonnage, is the weight of the water that a ship displaces when it is afloat. Standard displacement is usually used for naval vessels and is defined as the displacement of the ship, fully manned and equipped and ready for sea, but does not include fuel or reserve boiler feed water (which is why standard displacement is less than a ship's full load displacement).
11. Polmar, Norman, *Aircraft Carriers*, pp. 136–137.
12. Sigal, Edward B., "The Navy Distillate Fuel Conversion Program." The Navy had been burning Naval Special Fuel Oil (NFSO) since World War II. In 1954 new "D" type boilers, which operated at 1,200 psi, were introduced. These boilers had several

advantages over the 600 psi plants previously used, including greater efficiency, reduced weight and volume, simplicity of operation with ease of maintenance, and greater reliability. In the early 1970s the Navy switched to Naval Distillate Fuel for shipboard propulsion and electric generating systems.

13. Friedman, Norman, *U.S. Aircraft Carriers*, p. 258.
14. Friedman, Norman, *U.S. Aircraft Carriers*, p. 263.
15. Friedman, Norman, *U.S. Aircraft Carriers*, p. 258.
16. Friedman, Norman, *U.S. Aircraft Carriers*, p. 266.
17. Friedman, Norman, *U.S. Aircraft Carriers*, pp. 264–265.
18. Friedman, Norman, *U.S. Aircraft Carriers*, p. 380.
19. Friedman, Norman, *U.S. Aircraft Carriers*, p. 381.
20. Friedman, Norman, *U.S. Aircraft Carriers*, p. 265.
21. "USS *Forrestal* Celebrates Ten Years."
22. The commissioning pennant is a long streamer that is the distinguishing mark of a commissioned Navy ship. The American version is blue at the hoist with seven white stars, the rest being single strips of red and white. It is flown at all times as long as the ship is in commission, except when an embarked flag officer or civilian official flies his personal flag in its place. Crewmen who are original crew members are known as "plank owners" and often accorded special recognition.
23. MacDonald, Scot, *Evolution of Aircraft Carriers*, pp. 71–72; Friedman, Norman, *U.S. Aircraft Carriers*, p. 270.
24. The Suez Crisis was a diplomatic and military confrontation in late 1956 between Egypt on one side, and Britain, France, and Israel on the other when Egyptian president Nasser nationalized the Suez Canal, prompting Britain and France to intervene militarily while Israel invaded Egypt. The U.S., Soviet Union, and U.N. played major roles in forcing Britain, France, and Israel to withdraw.
25. MacDonald, Scot, *Evolution of Aircraft Carriers*, pp. 71–72.
26. "United States Naval Aviation 1910–1995," p. 218.

Chapter 2. Thunder Ballet

1. On 20 December 1963 Carrier Air Groups (CVG) on attack carriers were redesignated as Carrier Air Wings (CVW).
2. Friedman, Norman, *U.S. Aircraft Carriers,* p. 397.
3. In 1956 JP-5 became the standard fuel for all Navy jet aircraft, but high octane aviation gasoline (AVGAS) continued to be required on board carriers for all reciprocating engine aircraft until they were phased out of service.
4. "The US Navy Aircraft Carriers."
5. Thomason, Tommy H., "Catapult Development."
6. Friedman, Norman, *U.S. Aircraft Carriers,* p. 257.
7. "The US Navy Aircraft Carriers."
8. There are many types of upkeep periods. They differ in the length of time required and the extent of alterations and repairs. These include Restricted Availability (RAV); Selected Restricted Availability (SRA); Drydocking Selected Restricted Availability (DSRA); Planned Incremental Availability (PIA); and Complex Overhaul (COH).

Chapter 3. The Golden Age

1. The radical design of the Vought F7U Cutlass apparently had its origins in German jet fighter designs from World War II.
2. The F8U had other advanced features for its day, including an area ruled fuselage, all-moving stabilators, dog-tooth notching at the wing folds for improved yaw stability, and liberal use of titanium in the airframe.
3. Early F8Us had a retractable fuselage tray with 32 unguided 70mm "Mighty Mouse" Folding-Fin Aerial Rockets (FFAR), but these were rarely used and the tray was often sealed shut and was eliminated on later production aircraft.
4. That the Crusader was highly regarded by its pilots was reflected in the motto—"When you're out of F-8's, you're out of fighters."
5. The probe and drogue aerial refueling system was originally developed by the British firm of Flight Refueling, Limited.
6. "MK 5 Ejection Seats," Martin-Baker website; "United States Naval Aviation, 1910–1995," p. 219.
7. These carriers operated mostly in the Atlantic and Mediterranean in the late 1950s. In the Western Pacific, the smaller *Essex* attack carriers were assigned four-plane A3D detachments to give the Seventh Fleet a similar capability.
8. Before the operational debut of the Savage in 1951, an interim nuclear attack capability was provided by specially configured twin engine Lockheed PV-2 Neptune land-based patrol bombers, which could be craned on board the *Midway*-class carriers and launched on one-way nuclear attack missions.
9. Early model AJ-1 Savages were converted as tankers, by installing refueling equipment in the bomb bay with additional fuel in drop tanks. The later AJ-2 models were converted to the AJ-2P photographic reconnaissance version. It had 18 cameras and carried photo-flash bombs for night missions.
10. The Mark 7, the first nuclear weapon intended for carrying externally by fighter/attack aircraft, became available in January 1952.
11. The Skyhawk had some interesting characteristics from a carrier deck handling perspective. Besides not needing folding wings, blue-shirted aircraft handlers would hang from the scooter's wingtips while taxiing to help keep them from tipping over on their long spindly landing gear.
12. Airborne early warning began with Project Cadillac in World War II as an attempt to increase the range at which threat aircraft could be detected. By using an airborne radar that could transmit radar signals to the surface ships of a carrier task force the detection range could be increased to 100 miles. The TBM-3W Avenger with the AN/APS-20 radar entered service in March 1945.
13. The TF-1 Trader became the C-1, which served until 1988.
14. The Piasecki Helicopter Corporation, a pioneer in the development of tandem-rotor helicopters, changed its name in 1956 to Vertol Aircraft Corporation.
15. The Sparrow II was an active radar version that did not enter service.
16. The Vertol Aircraft Corporation (the former Piasecki Helicopter Corporation) was bought by Boeing Aircraft Company in 1960 and became Boeing-Vertol.
17. The various models of the Sparrow III became the AIM-7C, AIM-7D, and AIM-7E.

18. China Lake has a long history of coming up with simple solutions to ordnance development problems. The origin of the quote is uncertain, but aptly reflects the China Lake design philosophy.
19. "United States Naval Aviation, 1910–1995," p. 226.
20. Other countries still use the Bullpup as inert practice weapons.
21. The AGM-65 Maverick was introduced in 1972 and has been widely produced in a number of versions. It has the same form and dimensions as the AIM-4 Falcon and AIM-54 Phoenix.
22. The Rockeye II was so named because it followed an earlier Mark 12 Rockeye I that was canceled. It is technically the CBU-100 under the new system, but is usually still referred to as the Mark 20.

Chapter 4. The *Kitty Hawk*

1. The Key West Agreement of 1948 gave the Air Force the lead in developing intercontinental missiles, although the Army was allowed to continue development of its Jupiter missile. The Navy missile programs at the time were tactical weapons: the Loon, essentially a copy of the German V-1, and the Regulus. The Navy had no formal role in ballistic missile development until 1955, when the Special Projects Office under Rear Admiral William F. "Red" Rayborn was established. Isenberg, Michael T., *Shield of the Republic*, p. 659. Ten aircraft carriers were configured to operate Regulus missiles, but only six ever actually launched one. *Saratoga* did not deploy with the Regulus but was involved in two demonstration launches.
2. Polmar, Norman, *Aircraft Carriers*, p. 302.
3. In the early days of radar it was discovered that aircraft targets would disappear from radar screens only to reappear. These "fade zones" were due to atmospheric phenomena and the radar's characteristics. By charting when aircraft at known altitudes faded in and out approximate altitudes could be determined. These fade charts were used in World War II until specialized height finding radars were developed.
4. When CICs were introduced in the early 1940s, evolving from the primitive "Radar Plot" rooms, they were intended to be called Combat Direction Centers, but the name was changed when senior naval officers balked at being told what to do. CICs were later redesignated as CDCs.
5. Boslaugh, David L., CAPT USN (Ret.), "First-Hand: No Damned Computer is Going to Tell Me What to DO—The Story of the Naval Tactical Data System, NTDS."
6. The OPEVAL ships included guided-missile frigates *King* and *Mahan* as well as the *Oriskany*. They were later joined by the nuclear-powered carrier *Enterprise* and the cruiser *Long Beach*.
7. *America* was the first of her class to have NTDS, probably as commissioned in January 1965. She was followed by *Kitty Hawk* and *Constellation*. In August 1964 *Kitty Hawk* began an eight month yard period that included installation of Naval Tactical Data System (NTDS), Integrated Operational Intelligence Center (IOIC), Automatic Landing System (SPN-10), and the Airborne System Support Center (ASSC). *Constellation* had NTDS installed in 1965. In 1966–1967 she was followed by *Forrestal* at Norfolk, *Independence* at Newport, and *Ranger* at Bremerton.

8. The Soviet Navy continued to build and deploy diesel-electric attack submarines throughout the Cold War. The first Soviet ballistic missile submarines in the late 1950s were also diesel-electric, but by 1960, the Soviet Navy had launched its first nuclear-powered attack and ballistic missile submarines. Nuclear submarines designed to launch cruise missiles against American aircraft carrier task forces were designated SSGN by NATO. At its peak in 1980, the Soviet submarine force numbered 480 boats, including 71 fast attacks and 94 cruise and ballistic missile submarines. "Soviet Submarines."
9. The Sound Surveillance System (SOSUS) is a chain of underwater listening posts across the northern Atlantic Ocean near Greenland, Iceland, and the United Kingdom—the GIUK gap. Originally set up for tracking Soviet submarines coming west through the gap, other stations were added in the Atlantic and Pacific. SOSUS was later supplemented by the Surveillance Towed Array Sensor System (SURTASS) and became part of the Integrated Undersea Surveillance System (IUSS).
10. Friedman, Norman, "U.S. Aircraft Carrier Evolution: 1945–2011" in Douglas Smith's *One Hundred Years of U.S. Navy Airpower*, p. 337.
11. Polmar, Norman, *Aircraft Carriers*, p. 288.
12. Weitzenfeld, D.K., RADM, "Fleet Introduction of Colin Mitchell's Steam Catapult."
13. Friedman, Norman, *U.S. Aircraft Carriers*, pp. 278–279, 285.
14. In a beam riding system the missile flies down the guidance beam, which is pointed at the target. It was widely used on early missile systems, but has been largely replaced by semi-active systems for longer-range air-to-air and surface-to-air missiles. In a semi-active radar system, the missile has a detector that homes in on the radar signal reflected off the target from an external source.
15. *Enterprise* was eventually armed with 2 NATO Sea Sparrow launchers, 2 20mm Phalanx CIWS mounts, and 2 RAM launchers.
16. "At Sea With The Carriers," *Naval Aviation News*, January 1965, p. 33.
17. A shakedown cruise is a naval term for the at sea period performed before a ship enters service or after major changes occur, such as a crew turnover, repair, or overhaul. For new ships, the main reasons are to ensure everything is functioning and that the crew knows its job. A warship is usually not considered operational until it completes its shakedown cruise.
18. Friedman, Norman, *U.S. Aircraft Carriers*, p. 280.
19. The *Kitty Hawk* was ordered under the FY56 budget, the *Constellation* in FY57, and the *Enterprise* in FY58. *America* was under the FY61 budget and JFK in FY63.
20. Polmar, Norman, *Aircraft Carriers*, pp. 222–224.
21. The name will be carried on by the future *Gerald R. Ford*–class aircraft carrier *John F. Kennedy* (CVN-79).
22. The RIM-24 Tartar was a lightweight medium-range surface-to-air missile system designed for smaller ships that could engage targets at close range. Basically, it was the RIM-2C Terrier without the secondary booster. The other member of the "Three Ts," the RIM-8 Talos saw limited service because of its size and its dual radar antenna system, which few ships could accommodate.
23. Friedman, Norman, *U.S. Aircraft Carriers*, pp. 309–321.
24. Polmar, Norman, *Aircraft Carriers*, pp. 287–288.

Chapter 5. Cold Wars and Hot Spots

1. Polmar, Norman, *Aircraft Carriers*, p. 169.
2. Polmar, Norman, *Aircraft Carriers*, pp. 163–164, 168.
3. *Forrestal* returned to Norfolk 12 December to prepare for her first deployment to the Mediterranean, departing on 15 January 1957.
4. Polmar, Norman, *Aircraft Carriers*, pp. 149–152.
5. "United States Naval Aviation 1910–1995."
6. The crisis would continue until 1966 when Joseph Mobutu seized power with the help of the CIA. His regime would last for three decades and he changed the country's name to Zaire in 1971. With the ending of the Cold War in the early 1990s, western support evaporated and he was eventually deposed, dying in exile in Morocco in 1997.
7. Siad Barre had tried to seize the Ogaden region of Ethiopia, which by then had also become friendly with the Soviet Union. Somali forces were successful until a massive Soviet intervention by 20,000 Cuban forces and several thousand Soviet experts on behalf of Ethiopia's communist regime.
8. The *Liberty* (AGTR-5) was ostensibly a "technical research ship" that had been in position to assist in communications between American diplomatic posts in the region and to support the evacuation of American dependents, if necessary.
9. "Dictionary of American Naval Fighting Ships."
10. Polmar, Norman, *Aircraft Carriers*, pp. 152–153.
11. "United States Naval Aviation 1910–1995."
12. Polmar, Norman, *Aircraft Carriers*, p. 169.
13. Polmar, Norman, *Aircraft Carriers*, p. 213–216.
14. Faltum, Andrew, *The* Essex *Class Aircraft Carriers*, pp. 139–140.
15. Murphy, John, CDR USN (Ret.), "Cold War Warriors: Incidents at Sea."
16. Early information from the intelligence community identified this as the Tu-20. Although that was the initial Soviet designation, it had become the Tu-95 by the time it reached operational units.
17. Polmar, Norman, *Aircraft Carriers*, pp. 220–222.
18. Since the early Cold War, spy ships were used by all the major powers. In addition to listening in on communications and tracking an opponent's fleet movements, they monitored nuclear tests and missile launches as well as gathering submarine "signatures"—the distinctive noise patterns that could often identify specific types of submarines—that provided useful information for antisubmarine warfare. During that era, the U.S. had about 80 vessels, usually classified as "environmental research" vessels, while the Soviets had around 60 ships, often converted fishing trawlers or hydrographic research ships. The U.S. Navy terminology for these intelligence collection "trawlers" was AGI, which stood for "Auxiliary General Intelligence."
19. "Intrusions, Overflights, Shootdowns and Defections During the Cold War and Thereafter."
20. Murphy, John, CDR USN (Ret.), "Cold War Warriors: Incidents at Sea."
21. Many of the provisions of the treaty dealt with how warships of the two nations would avoid dangerous or provocative actions when operating near each other, such as not interfering with each other's formations or avoiding maneuvers in heavy sea

traffic. Prohibited actions included simulating attacks at, launching objects toward, or illuminating the bridges of the other party's ships. Other provisions dealt with the conduct of surveillance ships, submarines, and aircraft. Aircraft were to use caution in approaching aircraft and ships of the other party and not permitted to simulate attacks against aircraft or ships, perform aerobatics over ships, or drop hazardous objects near them.

22. Murphy, John, CDR USN (Ret.), "Cold War Warriors: Incidents at Sea."
23. "Dictionary of American Naval Fighting Ships."
24. Polmar, Norman, *Aircraft Carriers*, p. 222.

Chapter 6. Next-Generation Aircraft

1. The General Dynamics F-111 Aardvark was a medium-range tactical strike aircraft that first entered service in 1967. There were also strategic bomber, reconnaissance, and electronic warfare versions. The Royal Australian Air Force (RAAF) began operating F-111Cs in 1973. The F-111 pioneered variable-sweep wings, afterburning turbofan engines, and automated terrain-following radar for low-level, high-speed flight. Air Force F-111 variants were retired in the 1990s (F-111Fs in 1996 and EF-111s in 1998). RAAF F-111s served until 2010.
2. It was second only to the North American Sabre in numbers built (third if the T-33 trainer versions are included with the Lockheed F-80) and was produced under license in Japan. The Phantom is still in use by the Republic of Korea.
3. Although the FH-1 Phantom was the Navy's first pure jet fighter, it was preceded by the Ryan FR-1 Fireball, a piston- and jet-powered aircraft developed during World War II. Only 66 FR-1 aircraft were built before the end of the war and only one squadron flew the Fireball. The FR-1 was not sturdy enough for carrier operations and was withdrawn in 1947.
4. The F4H-1 was the initial production version of the Phantom. Since the J79-GE-8 engines intended for the Phantom were not yet available, production began with J79-GE-2 or -2A engines. To distinguish them from later models they were redesignated F4H-1F (the F indicated the use of a different powerplant). The F4H-1F became the F-4A in 1962.
5. The first true operational Phantom was the F-4B.
6. As the first deployable F-4 squadron VF-74 adopted the motto "First in Phantoms."
7. For a brief time the Air Force version was known as the F-110 Spectre. The Air Force version had dual controls and other changes, such as wider, lower pressure tires and anti-skid wheel brakes.
8. Polmar, Norman, *Aircraft Carriers*, p. 204.
9. 18 RA-5Cs were lost in combat: 14 to anti-aircraft fire, 3 to SAMs, and 1 to a MiG-21. Nine more were lost in operational accidents. 36 additional RA-5C aircraft were built from 1968 to 1970 as attrition replacements.
10. Later this would evolve into the Carrier Intelligence Center (CVIC).
11. The smaller *Essex*-class carrier had only three plane detachments of RF-8A/G Crusaders.

12. The DIANE system featured a display screen for the pilot that simulated the ground and sky and provided steering cues. By today's standards of computer-generated imagery it was extremely crude.
13. The TRAM turret only served to make the already homely Intruder even uglier. As a squadron intelligence officer with VA-115 on the *Midway*, the author used to refer to the A-6 as "an airplane that only a Bombardier/Navigator could love."
14. Polmar, Norman, *Aircraft Carriers*, p. 204. At one point the *Kennedy* had an experimental air wing with two A-6 squadrons.
15. Because it had a rudimentary attack system, the KA-6D could, in principle, carry offensive stores. In practice, however, this capability was rarely, if ever, used.
16. During World War II, under the code name Project Raven, specialists in operating equipment to detect, identify, and defeat enemy radars were officially known as Radar Observers, but more commonly called Ravens. In the U.S. military the term evolved to become Crows.
17. Polmar, Norman, *Aircraft Carriers*, p. 278.
18. Polmar, Norman, *Aircraft Carriers*, pp. 205–206.
19. These "re-procured" aircraft are designated C-2A(R) because they are so similar to the original aircraft, although they have airframe improvements and better avionics. The older C-2As were phased out in 1987, and the last of the new models was delivered in 1990.
20. Ling-Temco-Vought (LTV) was a large U.S. conglomerate that, at its peak, included aerospace, electronics, steel manufacturing, sporting goods, and airline industries as well as meat packing, car rentals, pharmaceuticals, and other businesses. At the time the A-7 was being developed, this included what had been the Chance Vought Corporation, which was acquired in 1962 and developed aircraft for the Air Force and Navy as LTV.
21. Baugher, Joseph F., "Ling-Temco-Vought A-7A."
22. The A-7C was a two-seat version intended as a trainer and the A-7D was the Air Force version with the more powerful Allison TF41 turbofan engine, a 20mm Vulcan gatling gun and different electronics. Early production A-7Es were equipped with TF-30-P-5 engines and were delivered as A-7Cs.

Chapter 7. New Developments

1. "USS Enterprise CVN 65," U.S. Carriers website.
2. Femiano, Don, "Look Ma, No Hands! The Automatic Carrier Landing System (ACLS)."
3. "Aviation Electronics Technician."
4. "A Brief History of U.S. Navy Aircraft Carriers: Part V—Space and Vietnam."
5. "Aircraft Carriers—CV/CVN," http://navysite.de/cruisebooks/cv60–71/004.htm.
6. Polmar, Norman, *Aircraft Carriers*, pp. 292–294.
7. "CV Concept (CCON) Study Report."
8. Friedman, Norman, *U.S. Aircraft Carriers*, p. 266.
9. The *Kitty Hawk* and the *Nimitz* were the first carriers to have the RAM systems installed in March 2002. William T. Baker and Mark Evans, "The Year in Review 2002," *NANews Year in Review 2002*, July–August 2003.

10. "Navy Fuel Specification Standardization."
11. "Extending the Lives of Carriers," OP-55 Briefing Point Paper.
12. Polmar, Norman, *Aircraft Carriers*, pp. 207–208.
13. Polmar, Norman, *Aircraft Carriers*, pp. 219–220.
14. Polmar, Norman, *Aircraft Carriers*, pp. 207–209, 219.

Chapter 8. Vietnam

1. Air America was established in 1950 and lasted until 1976. The author met a number of Air America pilots in April 1975 during Operation Frequent Wind, where they were instrumental in picking up evacuees from around Saigon and transferring them to the main evacuation sites.
2. "USS Constellation CV 64," U.S. Carriers.net.
3. "Guideline" was the NATO reporting name for the Soviet S-75 Dvina missile. Under this system the names of surface-to-air (or ground-to-air) missiles begin with the letter G.
4. "United States Naval Aviation 1910–1995."
5. "Dictionary of American Naval Fighting Ships."
6. "Dictionary of American Naval Fighting Ships."
7. The flares were magnesium parachute flares. Subsequent analysis showed that one in a thousand flares *could* ignite if jarred, and the cause was attributed to human error. The design of the flares was later changed to be immune to accidental ignition.

Chapter 9. The *Forrestal* Fire

1. The safety mechanism on the Zuni rocket had prevented it from detonating, but the impact had torn the tank off the wing and ignited the resulting spray of escaping JP-5 fuel, causing an instantaneous fireball. Within seconds, other external fuel tanks on White's aircraft overheated and ruptured, releasing more jet fuel to feed the flames, which began spreading along the flight deck.
2. McCain volunteered to stay in WestPac on board the *Oriskany* and became a POW when he was shot down on his 23rd mission on 26 October 1967. He remained a prisoner for the next five and a half years.
3. Of the nine bomb explosions that eventually occurred on the flight deck, eight were Composition B, the ninth was a Composition H6 bomb set off by a sympathetic detonation with an older bomb. Other ordnance, such as missiles, rockets and 20mm shells were also set off by the fires.
4. The Farrier Fire Fighting School Learning Site in Norfolk is named for Chief Farrier.
5. Quote from Rear Admiral Harvey P. Lanham, ComCarDiv Two.
6. "USS Forrestal CV 59," US Carriers.net.
7. These reviews still exist as the Weapon System Explosives Safety Review Board.
8. "At Sea With the Carriers: *Independence* (CVA-62)," *Naval Aviation News*, February 1968, p. 37. The *Independence* also tested a new "double probe" refueling device that used two fuel hoses for underway replenishment. This cut down the time required for the receiving ship to remain alongside the oiler and allowed almost instant breakaway without broken hoses.

Chapter 10. Turning Point and Aftermath

1. The other was a MiG-17 shot down by an F-8 from the *Oriskany* on 14 December 1967.
2. The Mark 80 series was developed in the 1950s in several nominal sizes: Mark 81 250 pounds; Mark 82 500 pounds; Mark 83 1,000 pounds; and Mark 84 2,000 pounds. The Mark 81 was deemed ineffective and was withdrawn from service, but has come back since experience in Iraq has shown the need to limit collateral damage. The Air Force developed a "ballute" air bag; it deploys from the tail to perform the same function.
3. "USS America CV 66," US Carriers.net.
4. "USS Constellation CV 64," US Carriers.net.
5. "United States Naval Aviation 1910–1995."
6. "USS Saratoga CV 60," US Carriers.net.
7. VF-103 began its career as the famous Corsair squadron (VF-17) in World War II. The Jolly Rogers have flown over nine different types of aircraft and their skull and crossbones insignia is one of the most widely recognized in the world. They currently serve as VFA-103 flying the FA-18.
8. This was the only all-Marine MiG kill. Two other Marines shot down MiGs but they were on exchange with the Air Force.
9. "United States Naval Aviation 1910–1995."
10. "Dictionary of American Naval Fighting Ships."
11. Stein, Stephen K., "Racial Unrest in the U.S. Navy, 1972–73."
12. Polmar, Norman, *Aircraft Carriers*, pp. 279–281.
13. "United States Naval Aviation 1910–1995."
14. The author was on the *Midway* during Frequent Wind as a squadron intelligence officer. The sights and sounds of many helicopters circling the ship and the controlled chaos on the flight deck as evacuees were led to safety below were to become indelible memories.
15. Between 1964 and 1975 each attack carrier had made at least one deployment to Southeast Asia: *America* 3; *Bon Homme Richard* 6; *Constellation* 8; *Coral Sea* 8; *Enterprise* 7; *Franklin D. Roosevelt* 1; *Forrestal* 1; *Hancock* 9; *Independence* 1; *Intrepid* 3; *Kitty Hawk* 9; *Midway* 9; *Oriskany* 10; *Ranger* 8; *Saratoga*; *Shangri-La* 1 and *Ticonderoga* 7. ASW carriers that participated included: *Bennington* 4; *Hornet* 3; *Kearsarge* 4; and *Yorktown* 3. Polmar, Norman, *Aircraft Carriers* p. 282.

Chapter 11. Evolution

1. While the F-111B did not enter service, land-based F-111 variants served with the Air Force for many years and with Australia until 2010.
2. Vice Admiral Thomas Connolly's testimony before Congress helped kill the F-111B. In responding to a question from Senator John C. Stennis as to whether a more powerful engine would cure the aircraft's woes, he stated, "There isn't enough power in all Christendom to make that airplane what we want!"
3. The Iranians, however, claimed a number of kills with the Phoenix against Iraqi aircraft during the Iran-Iraq war.

4. The Tomcat completed carrier trials on board the *Forrestal* at Norfolk in June 1972 using the new Mark 7 Jet Blast Deflectors that were installed to handle the heat blast of the TF30 engines.
5. The F-14 was sold to Iran in 1976 at a time when the U.S. had good diplomatic relations with the government of Shah Mohammad Reza Pahlavi. The surviving F-14s are now only in service with the Islamic Republic of Iran Air Force.
6. The Viking continues to serve in test and development roles.
7. Variants of the Sikorsky H-60 series are also in service with many foreign countries.
8. Variants of the Maverick have used electro-optical television, imaging infrared, laser imaging infrared, and charge-coupled device guidance. It continues in widespread service and has been exported to over 30 countries.
9. There is also the GBU-59 Enhanced Paveway II using the Mark 81 250-pound bomb.
10. The AGM-84 Harpoon SLAM can also be carried by the Air Force B-52.
11. During the Gulf War, the tail gunner of the B-52 had targeted an Air Force F-4G Wild Weasel, thinking it was an Iraqi MiG. The F-4 pilot launched the HARM missile and then saw that the target was the B-52, which was hit. It survived with shrapnel damage to the tail and no casualties. The B-52 was subsequently renamed "In HARM's Way."
12. "AGM-123 Skipper II," Federation of American Scientists.
13. "Joint Direct Attack Munition (JDAM)," Federation of American Scientists.
14. "AIM-120 AMRAAM," Federation of American Scientists.
15. The JSOW has been considered a very successful weapon system. In March 2003, for example, the *Kitty Hawk* used JSOWs during Operation Iraqi Freedom against Iraqi SAM sites. "Dictionary of American Naval Fighting Ships."

Chapter 12. Challenges

1. "Year in Review (1974)," *Naval Aviation News*, February 1975.
2. The *Belknap* was the lead ship of her class and had an aluminum superstructure, which melted during the fire. The Navy later came to the conclusion that the damage would have been less extensive if the superstructure had been made of steel and later classes would not have aluminum superstructures. *Belknap* would be rebuilt and recommissioned in 1980.
3. The author had the privilege of meeting one of the hostages, Colonel Leland J. Holland, when he spoke to our attaché training class.
4. There is also "The Great Gonzo" of Muppets fame, although any connection is unclear.
5. John F. Lehman has had a varied career, including military service in the Air Force and Navy reserves, work under Henry Kissinger on the National Security Council (1969–74), Secretary of the Navy in the Reagan administration, and as a member of the 9/11 Commission in 2003. As an advocate for the 600 ship Navy, he derided critics as "systems analysts," "armchair strategists," and "detentists." He has been quoted as saying "Power corrupts. Absolute power is kind of neat."

6. The SA-5 Gammon is the NATO reporting name for the Soviet S-200 Angara/Vega/Dubna very-long-range, medium-to-high altitude SAM system designed to defend large areas from bomber attack or other strategic aircraft. It entered service in 1967 and was exported to a number of countries, including Algeria, Iran, Kazakhstan, North Korea, and Syria. Libya received SA-5s from the Soviet Union in late 1985.
7. The Rockeye is a free-fall, unguided cluster bomb.
8. "USS Saratoga CV 60," US Carriers.net.
9. "USS John F. Kennedy CV 67," US Carriers.net.
10. Cooper, Tom, and Eric L. Palmer, "Disaster in Lebanon: US and French Operations in 1983."
11. Goldwater-Nichols is officially known as the Department of Defense Reorganization Act of 1986 Public Law 99–433.
12. "USS Saratoga CV 60," US Carriers.net.
13. *Thach* radioed the platforms. *Hoel*, *Leftwich*, *Kidd*, and *John Young* fired at the platforms. One was boarded by U.S. Special Forces, who recovered teletype messages and other documents, then planted explosives to destroy the platform. Air cover was also provided by *William H. Standley* as well as the *Ranger* aircraft.

Chapter 13. A Dangerous World

1. The term "joint" refers to operations between two or more services and "combined" refers to operations involving more than one nation, usually members of an alliance, such as NATO. "Coalition" operations often involve nations that are not part of a formal alliance, such as occurred during Desert Storm.
2. "U.S. Navy in Desert Shield/Desert Storm."
3. "United States Naval Aviation 1910–1995."
4. Saddam Hussein was found and captured in December 2003 and eventually tried and hanged in December 2006.
5. The SS-1 Scud is the NATO reporting name for a series of tactical ballistic missiles developed by the Soviet Union. The Soviet names include the R-11 for the first version and R-17 and R-300 Elbrus for later developments. The Scud has been widely exported to other countries, particularly in the Third World, and many countries besides Iraq have developed their own indigenously produced derivative versions. The term Scud has come into general use to describe such tactical ballistic missiles in general.
6. "U.S. Navy in Desert Shield/Desert Storm."
7. The rushing of American troops to the Persian Gulf region was under Operation Vigilant Warrior.
8. UNOSOM I was established in April 1992 and ran until its duties were assumed by the Unified Task Force (UNITAF) mission in December 1992. Following the dissolution of UNITAF in May 1993, the succeeding U.N. mission in Somalia was known as UNOSOM II.
9. The TCG *Muavenet* DM-357 was the former USS *Gwin* DM-33.
10. Since the damage was extensive, *Muavenet* was decommissioned. The *Knox*-class FFG-1093 USS *Capodanno* was later provided as compensation.
11. After Slovenia and Croatia seceded from the Socialist Federal Republic of Yugoslavia in 1991, the multi-ethnic Socialist Republic of Bosnia and Herzegovina passed a

referendum for *independence* in February 1992. The population was 44% Muslim Bosniaks, 31% Orthodox Serbs, and 17% Catholic Croats. The Bosnian Serbs had boycotted the referendum and established their own republic, supported by the Serbian government of Slobodan Milošević. The Yugoslav People's Army (JNA), mobilized forces inside the Republic of Bosnia and Herzegovina to secure Serbian territory, then war broke out across the country, accompanied by the ethnic cleansing of the Bosniak population, especially in Eastern Bosnia. The siege of Sarajevo and the Srebrenica massacre would become symbols of the conflict.

12. In February 1994 the crew of the *Saratoga* referred to its deployment in the Adriatic Sea, in support of Bosnia operations, as Groundhog Station. It was a reference to the 1993 movie *Groundhog Day* where the main character, played by Bill Murray, keeps living the same day over and over. In the military Groundhog Day has become synonymous with repetitive operations with no end in sight.
13. Land-based EP-3s provided intelligence collection. Navy air defense suppression capabilities were critical; the Air Force and Navy together provided more than 70% of NATO air defense suppression sorties during Deny Flight and nearly 60% during Deliberate Force. Navy tactical reconnaissance proved critical as well, because the mountainous terrain often masked areas to satellite coverage. Overall, Navy and Marine Corps aircraft flew over 13,500 sorties.
14. "USS America CV 66," US Carriers.net.
15. The conflict between the Taliban insurgency and the International Security Assistance Force (ISAF) is still going on. The Philippines and Indonesia, nations with their own internal struggles with Islamic terrorism, have also figured prominently in the war on terrorism. Operation Enduring Freedom is a joint U.S., U.K., and Afghan effort that is separate from the ISAF, a NATO effort that is run in parallel.
16. Vogel, Steve, "A Carrier's Quiet, Key Mission."
17. Following the terrorist attacks Operation Infinite Justice was announced and almost immediately renamed because of the reaction of Muslims—only Allah can deliver infinite justice.
18. "Dictionary of American Naval Fighting Ships."
19. "Dictionary of American Naval Fighting Ships."

Chapter 14. Into the Sunset

1. Polmar, Norman, *Aircraft Carriers*, pp. 287–288.
2. There are also Reserve Category D ships in the custody of the Inactive Fleet. Only berthing support is provided; they are not considered inactive but temporarily retained pending usage by the active force.
3. "USS Ranger CV 61," US Carriers.net.
4. "USS Ranger CV 61," US Carriers.net.
5. USS Ranger (CVA/CV-61) History and Memorial Website.
6. "Forrestal: CVB 59-CVA 59-CV 59-AVT 59," Haze Gray and Underway.
7. "Saratoga (CV 60), (ex-CVA 60), Multi-Purpose Aircraft Carrier," Naval Vessel Register.
8. "USS America (CV 66)," Unofficial U.S. Navy Site.
9. Quoted letter from Vice Chief of Naval Operations Admiral John B. Nathman.

10. "USS America (CV 66)," Unofficial U.S. Navy Site.
11. In October 1973 the *Midway* became the first carrier to be permanently forward deployed as the result of an accord reached between the U.S. and Japan in 1972.
12. In 2001, during a pre-deployment trial, *Kennedy* was found to be materially deficient, especially in air operations. Two catapults and three aircraft elevators were non-functional during inspection, and two boilers would not light. Her captain and two department heads were subsequently relieved.
13. Air Wing 5 was passed down from the *Midway* to the *Independence* to the *Kitty Hawk*, with the squadrons changing as newer aircraft replaced older aircraft.

Terms and Abbreviations

AAA	Antiaircraft artillery
AAM	Air-to-air missile
AFFF	Aqueous Film Forming Foam
Air Boss	Nickname for the air officer, head of the Air Department.
Air Group	The complement of aircraft on an aircraft carrier. Attack carrier air groups became air wings in 1962.
Air Wing	The complement of aircraft on an aircraft carrier.
Alpha Strike	Term from the Vietnam War for a carrier strike on a pinpoint target simultaneously by multiple aircraft.
ARG	Amphibious Ready Group
ASW	Antisubmarine warfare
ASWOC	Antisubmarine Warfare Operations Center
Bagel Station	An operating area off the eastern coast of Crete for carriers supporting operations in Lebanon.
BDA	Bomb damage assessment
Black Shoes	Non-aviation naval officers.
Bolter	A missed carrier arrested landing.
BPDMS	Basic Point Defense Missile System (NATO Sea Sparrow)
Break	The point in the landing pattern where individual aircraft in a flight "break" into the landing pattern to establish the proper interval.
Brown Shoes	Naval aviation officers.
CAG	Commander of the air group. The commander of the air wing is still called CAG.
Caliber	In smaller weapons, refers to the diameter of the bore in hundredths of an inch, e.g., a "fifty caliber" machine gun has a bore diameter of .50 inches. In larger naval weapons, caliber refers to the length of the barrel, e.g., a 5"/54 gun has a barrel length of 270 inches (the bore in inches times the caliber).

CAP	Combat Air Patrol
CarDiv	Carrier Division
CATCC	Carrier Air Traffic Control Center
CDC	Combat Direction Center
CIC	Combat Information Center
CIWS	Close-In Weapons System
CNO	Chief of Naval Operations
COD	Carrier Onboard Delivery
CSG	Carrier Strike Group; previously CVBG
CVA	Attack Aircraft Carrier
CVBG	Carrier Battle Group; previously Carrier Air Group (Large), replaced by Carrier Strike Group (CSG) in 2003
CVG	Carrier Air Group; designation effective 1948; replaced by Carrier Air Wing (CVW) on 20 December 1963
CVIC	Carrier Intelligence Center
CVS	Antisubmarine Warfare Aircraft Carrier
CVSG	Antisubmarine Air Group
CVW	Carrier Air Wing; previously CVG
CVWR	Reserve Carrier Air Wing
Displacement	The weight of a ship, which is equal to the weight it displaces, used to indicate the size of a warship. Merchant ship "tonnage" refers to a ship's volume, from the ancient word "tun," a cask used to store cargo.
Dixie Station	An operating area off the coast of South Vietnam in the Gulf of Tonkin for carriers supporting operations in the south.
DMZ	Demilitarized zone
DSRA	Drydocking Selected Restricted Availability
ECM	Electronic countermeasures
Feet Wet	Fighter direction brevity code to indicate that the aircraft had left land and was now over water. (Feet dry indicated crossing the shore from the sea.)
"G"	Unit force of gravity
Gonzo Station	An operating area in the Indian Ocean during the Iran hostage crisis and the tanker wars. It was an acronym for "Gulf of Oman Naval Zone of Operations."
Groundhog Station	An area in the Adriatic during operations in Bosnia. An area 90 miles north of the equator in the Indian Ocean was also known as "Groundhog Station."
HC	Helicopter Combat Support Squadron
HS	Helicopter Antisubmarine Squadron
HU	Helicopter Utility Squadron
IOIC	Integrated Operational Intelligence Center
JBD	Jet Blast Deflector

LANTIRN	Low Altitude Navigation and Targeting Infrared for Night
LHA	Amphibious Assault Ship; combined features of the Amphibious Assault Ship (LPH), Amphibious Transport Dock (LPD), Amphibious Cargo Ship (LKA), and Dock Landing Ship (LSD)
LHD	Landing Helicopter Dock (Follow on to LHA)
LSO	Landing signal officer
Mini-Boss	Nickname for the assistant air officer
MOVLAS	Manually operated visual landing aid system
NAVSEA	Naval Sea Systems Command
NDF	Naval Distillate Fuel
NSFO	Navy Special Fuel Oil
NTDS	Naval Tactical Data System
NVR	Naval Vessel Register
OLS	Optical Landing System
PLAT	Pilot Landing Aid Television
PriFly	Primary Flight Control
Purple K	Powder used in fighting fires
RAM	Rolling Airframe Missile
RAN	Reconnaissance/Attack Navigator
RAV	Restricted Availability
RCVW	Replacement Carrier Air Wing
RIMPAC	Rim of the Pacific Exercise, a large international naval exercise held biennially during June and July of even-numbered years off Hawaii.
RIO	Radar intercept officer
RVAH	Reconnaissance Attack Squadron
SAM	Surface-to-air missile
Scramble	Launch aircraft as soon as possible
SIOP	Single Integrated Operational Plan
SLEP	Service Life Extension Program
Sponson	A projecting structure from a ship's hull, often used to mount weapons
SRF	Ship Repair Facility
TARPS	Tactical Airborne Reconnaissance Pod System
UNREP	Underway replenishment
VA	Attack Squadron
VAAW	All-Weather Attack Squadron
VAH	Heavy Attack Squadron
VAP	Photographic Squadron
VAQ	Electronic Attack Squadron
VAW	Carrier Airborne Early Warning Squadron
VB	Bombing Squadron
VERTREP	Vertical replenishment

VF	Fighter Squadron
VFA	Strike Fighter Squadron (previously Attack Fighter Squadron, 1980–1983)
VFP	Light Photographic Squadron
VS	Antisubmarine Warfare Squadron (previously Scouting Squadron, became Sea Control Squadron in 1994)
Vultures Row	Area of the carrier's island where those not on watch can observe flight operations
Wave Off	Direction to discontinue a landing approach and go around
WestPac	The Western Pacific
Yankee Station	An operating area off the coast of North Vietnam for carriers striking the north

Bibliography

Sources

The information in this book was compiled from several sources. Norman Friedman's *U.S. Aircraft Carriers: An Illustrated Design History* provided much of the information about the design and technical characteristics of the *Forrestal* and *Kitty Hawk* classes. Norman Polmar's *Aircraft Carriers: A History of Carrier Aviation and Its Influence on World Events, Volume II, 1946–2006* was another major source as well as information available from the Naval History and Heritage Command. Online sources included Haze Gray and Underway, NavSource Online, and U.S. Carriers.net. Two other notable websites are the German Unofficial U.S. Navy website, which has digital copies of ship cruise books, and the Japanese Go Navy website, which features information about ship and air wing deployments.

"A Brief History of U.S. Navy Aircraft Carriers: Part V—Space and Vietnam," Americas Navy website, accessed January 2013, http://www.navy.mil/navydata/nav_legacy.asp?id=23

"A Brief History of USS *John F. Kennedy* (CV 67)," excerpt from decommissioning program, Department of the Navy, 2007.

"A Brief History of USS *Kitty Hawk* (CV 63)," excerpt from decommissioning program, Department of the Navy, 2009.

"Agreement Between the Government of The United States of America and the Government of The Union of Soviet Socialist Republics on the Prevention of Incidents On and Over the High Seas," U.S. Department of State website, accessed January 2013, http://www.state.gov/t/isn/4791.htm

"Aircraft Carrier Photo Index," NavSource Online, accessed various dates, http://www.navsource.org/archives/02idx.htm

"Aircraft Carriers—CV/CVN," Unofficial U.S. Navy Site, accessed various dates, http://navysite.de/carriers.htm

"At Sea With the Carriers: Independence (CVA-62)," *Naval Aviation News*, February 1968, p. 37.

"Aviation Electronics Technician," Integrated Publishing, Inc. website, accessed January 2013, http://navyaviation.tpub.com/14030/css/14030_208.htm

Baugher, Joseph F., "Ling-Temco-Vought A-7A," American Military Aircraft, last revised 12 December 2001, http://www.joebaugher.com/usattack/newa7_1.html

Boslaugh, David L., CAPT USN (Ret.), "First-Hand: No Damned Computer is Going to Tell Me What to DO—The Story of the Naval Tactical Data System, NTDS," IEEE Global History Network website, December 2012, http://www.ieeeghn.org/wiki/index.php/NO_DAMNED_COMPUTER_is_Going_to_Tell_Me_What_to_DO_-_The_Story_of_the_Naval_Tactical_Data_System,_NTDS

"Carrier and Air Wing Deployments (1991)," *Naval Aviation News*, July–August 1992.

"Carrier and Air Wing Deployments (1992)," *Naval Aviation News*, July–August 1993.

"Carrier and Air Wing Deployments 1993," *Naval Aviation News*, July–August 1994.

"Carrier and Air Wing Deployments 1994," *Naval Aviation News*, July–August 1995.

"Carrier and Air Wing Deployments, 1995," *Naval Aviation News*, July–August 1996.

"Carrier and Air Wing Deployments, 1996," *Naval Aviation News*, July–August 1997.

"Carrier and Air Wing Deployments, 1997," *Naval Aviation News*, July–August 1998.

"Carrier and Air Wing Deployments, 1998," *Naval Aviation News*, July–August 1999.

"Carrier and Air Wing Deployments, 1999," *Naval Aviation News*, July–August 2000.

Cooper, Tom, and Eric L Palmer, "Disaster in Lebanon: US and French Operations in 1983," Air Combat Information Group website, 26 September 2003, http://www.acig.org/artman/publish/article_278.shtml

"CV Concept (CCON) Study Report," Chief of Naval Operations, OP-96/cp, Ser 00437P96, 15 September 1971.

"Dictionary of American Naval Fighting Ships," Naval History and Heritage Command, accessed various dates, http://www.history.navy.mil/danfs/index.html.

"Extending the Lives of Carriers," OP-55 Briefing Point Paper, Department of the Navy, 31 October 1978.

Faltum, Andrew, *The* Essex *Class Aircraft Carriers*, Baltimore, MD: Nautical & Aviation Publishing Company of America, Inc., 1996.

Federation of American Scientists Military Analysis Network website, accessed various dates, http://www.fas.org

Femiano, Don, "Look Ma, No Hands! The Automatic Carrier Landing System (ACLS)," Textron Systems Retirees Association, Inc. website, accessed January 2013, http://www.tsretirees.org/0%20newsroom/memory/Femiano.pdf

"Forrestal: CVB 59-CVA 59-CV 59-AVT 59," Haze Gray and Underway, accessed January 2013, http://www.hazegray.org/navhist/carriers/us_super.htm#cva59

Freeman, Gregory A., *Sailors to the End: The Deadly Fire on the USS* Forrestal *and the Heroes Who Fought It*, New York, NY: Avon Books, 2002.

———, *Troubled Water: Race, Mutiny, and Bravery on the USS* Kitty Hawk, New York, NY: Palgrave Macmillan, 2009.

Friedman, Norman, *U.S. Aircraft Carriers, An Illustrated Design History*, Annapolis, MD: Naval Institute Press, 1983.

Grossnick, Roy A., *United States Naval Aviation, 1910–1995*, Washington, DC: Naval Historical Center, Department of the Navy, 1996.

Hill, Steven D., "First in Defense: A History of USS *Forrestal* (CVA/CV/AVT 59)," *Naval Aviation News*, November–December 1993.

———, "Super Sara: A History of USS *Saratoga* (CVA/CV 60)," *Naval Aviation News*, November–December 1994.

"Intrusions, Overflights, Shootdowns and Defections During the Cold War and Thereafter," last revised 17 September 2012, http://myplace.frontier.com/~anneled/ColdWar.html

Isenberg, Michael T., *Shield of the Republic: The United States Navy in an Era of Cold War and Violent Peace, Volume I, 1945–1962*, New York, NY: St. Martins Press, 1993.

"John F. Lehman," Notable Names Data Base website, accessed January 2013, http://www.nndb.com/people/230/000043101/

"Joint Direct Attack Munition (JDAM)," U.S. Navy Fact File, accessed January 2013, http://www.navy.mil/navydata/fact_display.asp?cid=2100&tid=400&ct=2

"Lest We Forget: DM-357 TCG Muavenet," *Bosphorus Naval News*, 3 October 2010, http://turkishnavy.net/2010/10/03/lest-we-forget-dm-357-tcg-muavenet-2/

MacDonald, Scot, *Evolution of Aircraft Carriers*, Washington, DC: Government Printing Office, Office of the Chief of Naval Operations, Department of the Navy, 1964.

Miller, Richard F., *A Carrier at War: On Board the USS* Kitty Hawk *in the Iraq War*, Washington, DC: Potomac Books, Inc., 2005.

"MK 5 Ejection Seats," Martin-Baker, accessed January 2013, http://www.martin-baker.com/products/ejection-seats/mk1-9/mk5

Murphy, John, CDR USN (Ret.), "Cold War Warriors: Incidents at Sea," *Emmitsburg News Journal* website, accessed January 2013, http://www.emmitsburg.net/archive_list/articles/misc/cww/2010/sea.htm

"Navy Fuel Specification Standardization," Interim report July 1986–May 1987, Defense Technical Information Center website, http://oai.dtic.mil/oai/oai?verb=getRecord&metadataPrefix=html&identifier=ADA248542

"One Final Cruise for Kitty Hawk," *Naval Aviation News*, November–December 2008.

Polmar, Norman, *Aircraft Carriers: A History of Carrier Aviation and Its Influence on World Events, Volume II, 1946–2006*, Washington, DC: Potomac Books, Inc., 2008.

"Saratoga (CV 60), (ex-CVA 60), Multi-Purpose Aircraft Carrier," Naval Vessel Register website, accessed January 2013, http://www.nvr.navy.mil/nvrships/details/CV60.htm

Sigal, Edward B., "The Navy Distillate Fuel Conversion Program," *Naval Engineers Journal* website, accessed January 2013, http://onlinelibrary.wiley.com/doi/10.1111/j.1559-3584.1971.tb03541.x/abstract

Smith, Douglas V., *One Hundred Years of U.S. Navy Airpower*, Annapolis, MD: Naval Institute Press, 2010.

"SOSUS: The 'Secret Weapon' of Underseas Surveillance," *Undersea Warfare*, Winter 2005 Vol. 7 No. 2, http://www.navy.mil/navydata/cno/n87/usw/issue_25/sosus2.htm

"Soviet Submarines," National Museum of American History website, Smithsonian Institution, http://americanhistory.si.edu/subs/const/anatomy/sovietsubs/index.html

Stein, Stephen K., "Racial Unrest in the U.S. Navy, 1972–73," review of *Black Sailor, White Navy: Racial Unrest in the Fleet during the Vietnam War Era*, by John Darrell Sherwood, New York: New York University Press, 2007, H-Net Reviews, published July 2008, http://www.h-net.org/reviews/showrev.php?id=14687

"The US Navy Aircraft Carriers," http://www.navy.mil/navydata/ships/carriers/rainbow.asp

Thomason, Tommy H., "Catapult Development," U.S. Navy Aircraft History website, 19 January 2011, http://thanlont.blogspot.com/2011/01/catapult-innovations.html

"United States Naval Aviation 1910–1995," Naval History and Heritage Command, accessed various dates, http://www.history.navy.mil/branches/usna1 910.htm

"U.S. Navy in Desert Shield/Desert Storm," Chief of Naval Operations, Ser OO/lU500179, 15 May 1991, http://www.history.navy.mil/wars/dstorm/index.html

"USS America CV 66," U.S. Carriers website, last modified 9 January 2013, http://www.uscarriers.net/cv66history.htm

"USS America (CV 66)," Unofficial U.S. Navy Site, accessed January 2013, http://navysite.de/cvn/cv66.htm#gencha

"USS Constellation CV 64," U.S. Carriers website, last modified 27 December 2010, http://www.uscarriers.net/cv64history.htm

"USS Enterprise CVN 65," U.S. Carriers website, last modified 9 January 2013, http://www.uscarriers.net/cvn65history.htm

"USS *Forrestal* Celebrates Ten Years," *Naval Aviation News*, February 1965, p. 17.

"USS Forrestal CV 59," U.S. Carriers website, last modified 9 December 2010, http://www.uscarriers.net/cv59history.htm

"USS John F. Kennedy CV 67," U.S. Carriers website, last modified 9 December 2010, http://www.uscarriers.net/cv67history.htm

"USS Ranger CV 61," U.S. Carriers website, last modified 9 December 2010, http://www.uscarriers.net/cv61history.htm

USS Ranger (CVA/CV-61) History and Memorial Website, accessed January 2013, http://uss-ranger.org/

"USS Saratoga CV 60," U.S. Carriers website, last modified 9 December 2010, http://www.uscarriers.net/cv60history.htm

Vogel, Steve, "A Carrier's Quiet, Key Mission: Kitty Hawk Heads Home After Hosting Special Forces," *Washington Post*, 24 December 2001.

Weitzenfeld, D.K., RADM, "Fleet Introduction of Colin Mitchell's Steam Catapult," Royal Academy of Engineering website, accessed January 2013, http://www.raeng.org.uk/prizes/mitchell/pdf/Fleet_Introduction_Article.pdf

"World Aircraft Carriers List: Master List of US Carriers," Haze Gray and Underway, accessed various dates, http://www.hazegray.org/navhist/carriers/us_index.htm

Index

About the Author

Andrew Faltum served as an air intelligence officer on USS *Midway* home ported in Yokosuka, Japan, before joining the Naval Reserve, retiring as a commander. He has been an intelligence specialist for the Army Materiel Command, as well as a defense contractor providing analytical support to the Joint Staff.

The Naval Institute Press is the book-publishing arm of the U.S. Naval Institute, a private, nonprofit, membership society for sea service professionals and others who share an interest in naval and maritime affairs. Established in 1873 at the U.S. Naval Academy in Annapolis, Maryland, where its offices remain today, the Naval Institute has members worldwide.

Members of the Naval Institute support the education programs of the society and receive the influential monthly magazine *Proceedings* or the colorful bimonthly magazine *Naval History* and discounts on fine nautical prints and on ship and aircraft photos. They also have access to the transcripts of the Institute's Oral History Program and get discounted admission to any of the Institute-sponsored seminars offered around the country.

The Naval Institute's book-publishing program, begun in 1898 with basic guides to naval practices, has broadened its scope to include books of more general interest. Now the Naval Institute Press publishes about seventy titles each year, ranging from how-to books on boating and navigation to battle histories, biographies, ship and aircraft guides, and novels. Institute members receive significant discounts on the Press's more than eight hundred books in print.

Full-time students are eligible for special half-price membership rates. Life memberships are also available.

For a free catalog describing Naval Institute Press books currently available, and for further information about joining the U.S. Naval Institute, please write to:

Member Services
U.S. Naval Institute
291 Wood Road
Annapolis, MD 21402-5034
Telephone: (800) 233-8764
Fax: (410) 571-1703
Web address: www.usni.org

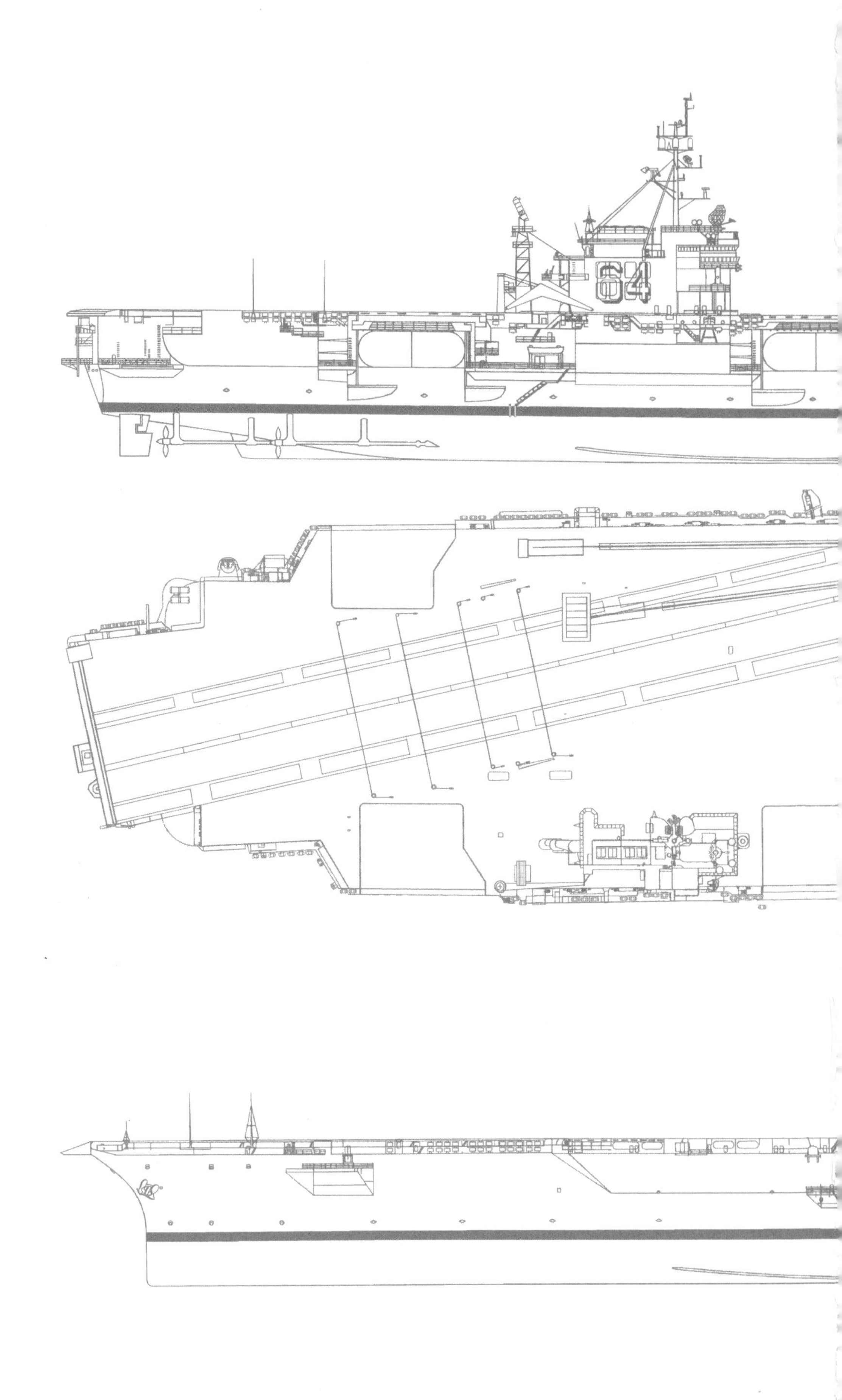
64